PROTEIN BIOCHEMISTRY, SYNTHESIS, STRUCTURE AND CELLULAR FUNCTIONS

RIBOSOMES

MOLECULAR STRUCTURE, ROLE IN BIOLOGICAL FUNCTIONS AND IMPLICATIONS FOR GENETIC DISEASES

PROTEIN BIOCHEMISTRY, SYNTHESIS, STRUCTURE AND CELLULAR FUNCTIONS

Additional books in this series can be found on Nova's website under the Series tab.

Additional e-books in this series can be found on Nova's website under the e-book tab.

MICROBIOLOGY RESEARCH ADVANCES

Additional books in this series can be found on Nova's website under the Series tab.

Additional e-books in this series can be found on Nova's website under the e-book tab.

PROTEIN BIOCHEMISTRY, SYNTHESIS, STRUCTURE AND CELLULAR FUNCTIONS

RIBOSOMES

MOLECULAR STRUCTURE, ROLE IN BIOLOGICAL FUNCTIONS AND IMPLICATIONS FOR GENETIC DISEASES

ZHOU LIN
AND
WANG LIU
EDITORS

NOVA BIOMEDICAL

New York

For permission to use material from this book please contact us:
Telephone 631-231-7269; Fax 631-231-8175
Web Site: http://www.novapublishers.com

Additional color graphics may be available in the e-book version of this book.

Library of Congress Cataloging-in-Publication Data

Library of Congress Control Number: 2012955268
ISBN: 978-1-62417-698-2

Published by Nova Science Publishers, Inc. † New York

Contents

Preface

Ribosomes are essential protein-RNA complexes that synthesize proteins according to the gene library of the cell in all living organisms. In this book, the authors present current research in the study of the molecular structure, biological functions and implications for genetic disease of ribosomes. Topics include ribosomal alterations and immune-related illness; the role of ribosomes in mediating hypoxia response and tolerance in eukaryotes; the potential biomedical use of ribosome-inactivating proteins; coarse-graining the nano-machine ribosome to elucidate its functional dynamics; and structural aspects of ribosomal ligands interactions providing the work of the human translational machinery.

Chapter 1 - The chapter is devoted to the review of structural principles of molecular processes providing functioning of the protein synthesizing system in higher organisms. Site-directed cross-linking and related data on structural elements of the human ribosome that form binding sites of mRNA, acceptor terminus of tRNA and hepatitis C virus IRES, and on peptides of the translation termination factor eRF1 participating in the organization of the stop codon recognition site are considered. Comparison of the structural information gained with translational system from higher organisms with the respective information on bacterial and lower eukaryotic systems is given. New insights into molecular mechanisms of translation based on the discussed data are presented.

Chapter 2 - Specific ribosome-directed stressors have the capacity to damage 28S ribosomal RNA and interfere with ribosomal functions during translation and cellular homeostasis. This can provoke so-called ribosomal stress responses that are closely associated with various disease processes including immunosuppressive or hypersensitivity-linked dysfunctions in humans and domestic animals. Since the primary toxic actions of most ribosomal stress-inducing agents generally inhibit global protein synthesis, highly dividing cells, including leukocytes and epithelial cells, are the most susceptible targets. Here, responses of immune cells and systems to acute and chronic exposure to ribosome-inactivating stress in various experimental models are discussed. This review will specifically focus on modulation of immune suppression and hypersensitivity diseases caused by ribosomal dysfunction. Evidence-based approaches characterizing ribosome-inactivating stressor and their implication as critical etiological factors of immune-associated diseases are also presented.

Chapter 3 - Ribosome biogenesis and protein synthesis are highly energy-consuming processes and therefore, key control points for cellular growth and division under various conditions of stress. Hypoxia is a state of poor oxygen supply that is experienced by diverse organisms under numerous physiological and disease conditions. Hypoxia greatly reduces cellular energy production and hence, organisms must develop effective strategies to cope with this low oxygen state. One common strategy that eukaryotes, ranging from yeast to mammals, employ to adapt under hypoxia is the conservation of cellular energy via suppression of metabolism and protein synthesis. These biological changes occur as a result of coordinated cellular responses orchestrated by various signal transducers and regulators, such as HIFs in higher eukaryotes and Mga2 and Hap1 in yeast. Studies in various eukaryotes have indicated that ribosomes have an essential role in mediating responses to hypoxia. In yeast the expression of various ribosomal and related proteins are altered under hypoxic conditions. Interestingly, it appears that deletion of certain genes involved in ribosome biogenesis, including *NSR1, BUD21, RAI1* or *RPL20A*, minimizes such alterations and enhances the survival of yeast under hypoxic conditions. Such deletions also diminish the effect of key transcriptional regulators, like Swi2 and Swi3, whose targets are altered under hypoxia in the wild-type strain. Similarly, in *C.elegans* reduced activity of arginyl-transfer RNA (t-RNA) synthetase confers hypoxia tolerance. Mammals acquire hypoxia tolerance via global reduction in RNA translation and ribosome biogenesis. In this chapter, the authors will review the role of ribosomes in conferring tolerance to hypoxia and discuss future directions to elucidate the molecular mechanisms underlying hypoxia tolerance.

Chapter 4 - Analyses suggest that protein coding genes occur within ribosomal RNA genes, enabling to code for additional proteins without increasing genome size. Stop codons can be translated by antisense antitermination (suppressor) tRNAs with anticodons matching stop codons, hence overlap coding might not be limited to stopless 'open reading frames'. Alignment analyses detect several putative protein coding genes within mitochondrial rRNAs and the mitochondrial D-loop region that include stops. These putative overlapping genes are probably functional because corresponding RNA abundances in GenBank's EST database increase with the length of the putative overlap coding region (but not for regions not involved in overlap coding), and predicted protein secondary structures resemble those formed by their presumed homologues from GenBank. In addition, numbers of putative overlap coding genes increase with predicted capacities of antisense tRNAs to match stop codons when comparing mitochondrial genomes from each primate and *Drosophila* species. In human mitochondrial genomes, single nucleotide polymorphisms are less frequent in putative rRNA-encoded overlapping protein coding genes than in the rest of each 12S, 16S rRNAs and the D-loop, as one would expect if the coding function of these regions cause excess purifying selection against polymorphisms. However, further analyses based on polymorphism frequencies and proportions of pathogenic mutations among these yield partially inconclusive results, probably because of weaknesses intrinsic to indirect estimation of purifying selection as indirect evidence for gene function. Previous analyses showed that overlapping protein coding genes in frameshifted and antisense sequences of regular mitochondrial protein coding genes coevolve with capacities of antisense tRNAs to translate stop codons, reassigning stop codons to code for amino acids: UAR are reassigned to glutamine in primates, tryptophane in turtles, and serine in *Drosophila*; AGR stops are reassigned to arginine in primates and glycine in the olive ridley (*Lepidochelys*) and lysine in other turtles. Analyses of mitochondrial rRNAs converge with analyses of protein coding genes and strengthen the hypothesis that mitochondrial genomes code according to a stopless genetic code for previously undetected, cryptic overlapping protein coding genes. It is probable that pathogenicity of at least some polymorphisms is due to the previously unsuspected function of rRNA and D-loop regions as protein coding regions.

Chapter 5 - Temperature up-shift or down-shift is one of the major environmental stresses encountered by microorganisms. Especially, one of the main consequences of temperature down-shift on cells is conformational changes of secondary structures of RNAs. These cold-shock induced structural changes of RNAs have significant effect on the global gene expressions, i) by improper termination of transcription, ii) inefficient RNA degradation, iii) translation of alternatively-structured mRNAs, and iv) biogenesis of ribosome and processing of rRNA. Several cold-shock proteins are produced to counteract these effects and thus allow cold acclimatization of the cell. This chapter describes how specific cold shock proteins modulate the ribosome biogenesis and eventually translation, to ensure effective protein production at low temperature.

Chapter 6 - Ribosome is crucial for protein synthesis in cells. Ribosome-inactivation proteins (RIPs) are a large family of unique enzymes that inhibit protein synthesis by site-specifically and irreversibly modifying the ribosomal RNA. According to their different enzymatic activity, RIPs can be classified into ribonuclease RIPs and N-glycosidase RIPs. RIPs have drawn the attention of many researchers due to their potential use as antivirus and anticancer reagents and as other biological and biomedical tools. In this chapter, the progress in the study of RIPs and the problems encountered in biological and biomedical research are discussed.

Chapter 7 - Coarse-grained computational techniques can provide information on conformational dynamics of the supramolecule ribosome, for which full-atom approaches become computationally demanding. Elastic network models and coarse-grained molecular dynamics simulations are the most commonly used methods for simple descriptions of the biomolecular structure. Usually, an amino acid or a nucleotide residue is represented by a single node, and simplified potential energy functions are used for interacting nodes. These approaches have been successful in reproducing the micro-to-milliseconds dynamics information, such as the experimentally observed ratchet-like rotation of the ribosomal subunits. Furthermore, they have provided insights on the structure-dynamics-function relationship of the ribosomal complex. This review summarizes various coarse-graining approaches on the nano-machine and recent studies that focus on its different functional sites.

Chapter 8 - In living cells and organisms, proteins are often made through decoding genetic information stored as sequences of nucleotides. The central machinery in the process is ribosome. Every protein is translated from the 5' end to the 3' end of an mRNA. Starting from the translation initiation site, every three consecutive nucleotides are considered as a codon and read as a single amino acid. Cellular proteins, which are made from 61 sense codons of a total 64 codons, typically comprise 20 natural amino acids. However, there are unusual cases existing in nature. Under certain circumstances, nature can use codons that are typically nonsense to code natural as well as unusual amino acids (*e.g.* selenocysteine and pyrrolysine). In addition, ribosomal frameshifting is also not uncommon, in which case one or two nucleotides in a coding sequence are skipped to make proteins based on a shifted frame. In the past decade or so, researchers have invented procedures in laboratories to biosynthesize proteins containing unusual amino acids. Such exploration has made a big impact. This chapter summarizes methods used in nature and in laboratories to incorporate unusual amino acids into proteins. In particular, recent studies, which re-engineered ribosome and other related translational factors, are emphasized. Therapeutic applications of this technology are also discussed.

Chapter 9 - Bowen-Conradi syndrome (BCS), a lethal autosomal recessive disorder which affects infants in the Hutterite population, is caused by a point mutation in *EMG1*, a highly conserved gene necessary for biogenesis of the small subunit of the ribosome. Research on EMG1 function has centered on yeast models, due to their genetic tractability for examining a cell-essential protein. The effects of the BCS-causing mutation on human EMG1 protein function are therefore unknown. The authors previously showed that EMG1 protein levels in BCS patient fibroblasts were severely reduced compared to control cells, and the authors therefore went on to examine EMG1 stability. Using pulse-chase metabolic labeling in cells transiently transfected with wild type or mutant EMG1, the authors found that the mutant protein rapidly associated with a detergent-insoluble fraction. Inhibition of the proteasome resulted in the accumulation of a higher-molecular weight band likely corresponding to ubiquitinated EMG1, suggesting that EMG1 is normally degraded by this pathway. Using immunofluorescence in fibroblasts, the authors found that both wild type and mutated EMG1 were localized to the nucleolus, and that proteasome inhibition resulted in the accumulation of EMG1 in nucleoplasmic foci which may correspond to sites of proteasome-mediated degradation. These data suggest that the mutated EMG1 protein is unstable, leading to a reduced level of EMG1 in the cell. Although the mutated protein localizes normally, the reduction in its level is likely to impact downstream pathways that significantly alter cell function in BCS patients.

In: Ribosomes
Editors: Zhou Lin and Wang Liu

ISBN: 978-1-62417-698-2

Chapter 1

Structural Aspects of Ribosomal Ligands Interactions Providing the Work of the Human Translational Machinery

Dmitri Graifer and Galina Karpova*
Laboratory of Ribosome Structure and Functions,
Institute of Chemical Biology and Fundamental Medicine,
Siberian Branch of the Russian Academy of Sciences, Novosibirsk, Russia

Abstract

The chapter is devoted to the review of structural principles of molecular processes providing functioning of the protein synthesizing system in higher organisms. Site-directed cross-linking and related data on structural elements of the human ribosome that form binding sites of mRNA, acceptor terminus of tRNA and hepatitis C virus IRES, and on peptides of the translation termination factor eRF1 participating in the organization of the stop codon recognition site are considered. Comparison of the structural information gained with translational system from higher organisms with the respective information on bacterial and lower eukaryotic systems is given. New insights into molecular mechanisms of translation based on the discussed data are presented.

Introduction

One of the key events of the cell life is protein synthesis that is conducted on ribosomes, very complicated ribonucleoproteins, whose function is translation of genetic information incoming as sequences of trinucleotides-codons of mRNAs into the polypeptide chains of proteins. Understanding molecular mechanisms of translation is the problem of basic

* E-mail: karpova@niboch.nsc.ru.

importance, especially with human ribosomes, because the nature of many diseases relates to disturbances in the functioning and regulation of the protein-synthesizing system.

The work of ribosomes has features common to all organisms. In the course of translation, the ribosome binds various specific ligands, which provides the sequence of events leading to the polypeptide chain synthesis. The main ligands of the ribosome are mRNAs bearing the genetic information to be translated, and transfer RNAs (tRNAs) carrying amino acid (aa) residues attached to the 3'-terminal riboses via complex ester bonds; each tRNA is specific to one particular aa. The mRNA binding site of the ribosome is located on the small subunit, and three tRNA binding sites (A, P and E sites) are composed of both small and large subunits. Aminoacyl (A) site is designed for binding of incoming aminoacyl-tRNA (aa-tRNA) whose anticodon recognizes mRNA codon bound at this site, i.e., in the A site selection of aa-tRNA cognate to the mRNA codon located at this site takes place. This process is called decoding. The function of the peptidyl (P) site is binding of tRNA carrying nascent peptide chain, and the role of the exit (E) site is binding of discharged tRNA before it leaves the ribosome. When the A site is occupied with aa-tRNA cognate to the mRNA codon present there, peptide bond between the peptidyl residue of the peptidyl-tRNA at the P site and aa residue of the aa-tRNA is immediately formed. Peptidyl transferase center (PTC) located at the large ribosomal subunit is responsible for the catalysis of peptide bond formation (transpeptidation). As the result, the peptidyl moiety elongated by one aa becomes attached to the A site tRNA, and the P site tRNA gets discharged. After the transpeptidation, the translocation takes place that includes the movement of the peptidyl-tRNA from the A to the P site, and of the discharged tRNA from the P to the E site with the accompanying movement of mRNA codons bound to the respective tRNAs. After completion of this process the ribosome becomes ready to the next peptide elongation cycle. The first elongation cycle is preceded by initiation of translation that is completed by formation of the ternary complex of the ribosome with mRNA and initiator tRNA aminoacylated with methionine (in bacteria, with formylmethionine), in which the mRNA start codon (usually, AUG) and Met-$tRNA_i^{Met}$ are bound at the P site. Termination of protein synthesis takes place when one of three triplets, UAA, UAG or UGA, binds at the A site. These triplets do not encode any aa and serve as stop signals (stop, or nonsense codons). The work of the ribosome is assisted with specialized proteins called translation factors. The factors operating in the elongation are functionally similar in all kingdoms of life; one of them (in prokaryotes, EF-Tu, and in eukaryotes, eEF-1) in complex with GTP delivers aa-tRNA to the A site and another (in prokaryotes, EF-G, in eukaryotes, eEF-2) promotes translocation in a GTP-dependent manner. Functional resemblance is also observed with factors taking part in translation termination in all organisms. So, stop codons are always recognized by class 1 termination factors, or polypeptide chain release factors (RF1/RF2 in bacteria and eRF1 in eukaryotes), that trigger hydrolysis of the complex ester bond in the peptidyl-tRNA, enabling the synthesized polypeptide to leave the ribosome; termination involves also class 2 termination factors that are ribosome-activated GTPases (RF3 in bacteria and eRF3 in eukaryotes).

On the other hand, the functioning of ribosomes from different kingdoms of life has a number of distinctions, which are the most pronounced between bacteria and eukaryotes. These distinctions concern mainly initiation and termination of translation. So, initiation in eukaryotes is much more complicated than in bacteria and involves more steps and larger set of initiation factors; moreover, the ways by which the mRNA start codon is placed at the P site in bacterial and eukaryotic ribosomes are principally different (for review, see Rodnina &

Wintermeyer, 2009; Voigts-Hoffmann et al., 2012; Aitken & Lorsch, 2012). In bacteria, mRNAs bind directly to small ribosomal subunits so that the AUG codon occurs at the P site, which is assisted by the complementary interactions between so called Shine–Dalgarno (SD) sequences, which are generally present in the mRNAs upstream of the start codon, and the anti-SD sequences near the 3'-terminus of the small subunit rRNA. Eukaryotic mRNAs lack SD sequences; they bind the small ribosomal subunit (that is already bound with initiator Met-tRNA) near the 5'-terminal 7-methylguanosine ("cap"), and then the subunit scans along the mRNA until the start codon AUG is recognized. The majority of eukaryotic initiation factors, including all ones responsible for the cap recognition and scanning, have no bacterial counterparts. The alternative pathway of translation initiation concerns genomic RNAs (gRNAs) of viruses, which have specific structured fragments upstream of the start codon, so called Internal Ribosomal Entry Site (IRES) that provide prokaryotic-like mode of initiation of their translation, namely, direct placement of the start codon at the P site region without scanning (e.g., see Hellen & Sarnow, 2001). Such IRES is contained in the gRNA of hepatitis C virus (HCV), one of the most dangerous human pathogens; this IRES is the object of numerous studies. As for the differences in translation termination in bacteria and eukaryotes, they concern, at first, the number of termination factors involved and the mechanisms of termination. Termination in bacteria is performed with two class 1 release factors, RF1 and RF2, which recognize UAG/UAA and UAA/UGA stop codons, respectively. In contrast, in eukaryotes all three stop codons are recognized with the same factor eRF1. Besides, in bacteria and eukaryotes roles of class 2 release factors and GTP hydrolysis in termination are different, as well as are mechanisms of GDP/GTP exchange on the RF3 and eRF3 (e.g., see Rodnina & Wintermeyer, 2009; Korostelev, 2011). Evidently, specific features of the work of eukaryotic translational machinery are based on molecular interactions that do not occur in bacteria.

The translational machinery became a subject of extensive studies utilizing various approaches in two last decades of XX century (reviewed in Rheinberger, 2004), but a breakthrough took place at the boundary between the XX and XXI centuries when the structures of bacterial ribosomal subunits had been deciphered at the atomic level by X-ray crystallography (Ban et al., 2000; Wimberly et al., 2000; Schluenzen et al., 2000). The Nobel Prize in Chemistry has been awarded for this work in 2009. Data on the structures of bacterial ribosomes and their various complexes with ribosomal ligands (tRNAs, mRNAs, etc.) obtained by means of X-ray crystallography made it possible to understand how the ligands interact with the ribosome and why the interactions are specific.

Eukaryotic ribosomes have a pronounced structural similarity with bacterial ones, but they are larger and more complex. So, eukaryotic rRNAs are longer than prokaryotic, and eukaryotic ribosomal subunits in addition to the complete set of ribosomal proteins homologous to their bacterial counterparts contain a number of other ribosomal proteins (about one-third of the total number). Ribosomes from eukaryotes still remain much less studied than ribosomes from bacteria. So, cryo-electron microscopy (cryo-EM) images of eukaryotic ribosomes reported (e.g., see Chandramouli et al., 2008; Taylor et al., 2009; Armache et al., 2010) do not provide direct information on the fine structures of the keystone functional sites of the ribosome, namely, of the mRNA binding center including the decoding site and of the peptidyl transferase center, since particular mRNA codons as well as tRNA acceptor ends were not yet visualized by this method. Very recently reported cryo-EM study of the mammalian ribosomal termination complex carried out at 17-18 Å resolution (Taylor et al., 2012) presented a mechanistic insight into the coordination between eRF3-dependent GTP hydrolysis and subsequent peptide release by eRF1 but did not provide information on amino acid fragments of eRF1 participating in the organization of stop codon recognition site. As for X-ray crystallography, the first recently reported studies concern only mRNA- and tRNA-free ribosomes from lower eukaryotes, namely, yeast (Ben-Shem et al., 2010; 2011) and *Tetrahymena thermophila* (Rabl et al., 2011; Klinge et al., 2011). Numerous data on the structural and functional organization of the human translational system have been accumulated with the use of a site directed cross-linking. This approach is based on the application of tRNA and mRNA analogs containing chemically reactive (generally, photoactivatable) groups in the selected locations. These ligands being bound within a model ribosomal complex form cross-links with neighboring components of the translational machinery, whose identification enables making conclusions regarding the structure of the ligand binding site.

This chapter is a review on site-directed cross-linking and other related data on the structures of the mRNA binding and the peptidyl transferase centers of the human ribosome and of the stop codon recognition site on the human eRF1, and on the structure of the HCV IRES binding site on the small ribosomal subunit. New insights into molecular mechanisms of translation in higher organisms based on the reviewed data are discussed.

Tools and Model Complexes

The set of tools for structural and functional studies of the human translational system included reactive derivatives of ribosomal ligands, namely, mRNA analogs, tRNA analogs and HCV IRES derivatives. Two types of mRNA analogs were used, namely, derivatives of relatively short (6 to 12 nt long) oligoribonucleotides bearing a *p*-azidotetrafluorobenzoyl (ATB) group attached to the base of the selected nucleotide (Demeshkina et al., 2000; 2003a & b; Graifer et al., 2004; Molotkov et al., 2005; 2006), or to the 3'- or the 5'-terminal phosphate (Bulygin et al., 1997; 2002; 2011) (Figure 1), and 40-50 nt long synthetic mRNAs carrying 4-thiouridine (s^4U) residues at designed positions (Chavatte et al., 2001; Bulygin et al., 2005; 2009).

HCV IRES derivatives bore either ATB groups at selected locations (Laletina et al., 2006; Babaylova et al., 2009) (Figure 2) or s^4U residues randomly substituted uridines over the whole molecule (Otto et al., 2002). tRNA analogs were used that contained s^4U either attached to the 3'-terminus of tRNA as an additional nucleotide (Bulygin et al., 2008) or introduced instead of the 3'-terminal A76 or the penultimate C75 (Bulygin et al., 2012). Chemical groups introduced into the ribosomal ligands can be activated with mild UV-light to form active intermediates capable of cross-linking to both RNA and proteins. In addition to photoactivatable tRNA analogs, tRNA derivatives with periodate oxidated 3' ribose (2',3'-dialdehyde derivatives of tRNA) were applied (Baouz et al., 2009; Hountondji et al., 2012). The latter derivatives were capable of cross-linking only to ribosomal proteins.

In cross-linking studies, model complexes were assembled by direct binding of the ligands to the ribosome without translation factors. In these complexes tRNA was bound at the P site (to which its affinity is higher than to the other ribosomal sites (see Watanabe, 1972; Lill et al., 1986; Graifer et al., 1992)) where it interacted with the cognate mRNA codon (Figure 3). To obtain complexes where, in addition to the P site, either the A or the E site was occupied with tRNA molecule too, tRNA cognate to the codon present at the respective site was bound to the ternary complex of 80S ribosomes with the mRNA analog and tRNA at the P site. By analogy, termination complexes were formed by binding of eRF1 to the ternary complexes of 80S ribosomes with tRNA at the P site and mRNA analogs, where the A site was programmed with a stop codon (Chavatte et al., 2001; 2002; Bulygin et al., 2002; 2003; 2010; 2011) (Figure 3, complex 5). In the related studies carried out with rabbit ribosomes, s4U-containing mRNA analogs were applied for site-directed cross-linking in 48S and 80S initiation complexes assembled with the use of the appropriate initiation factors (Pisarev et al., 2006; 2008). The cross-linking of HCV IRES derivatives was studied in their binary complexes with 40S ribosomal subunits, which are easily formed without any initiation factors (Pestova et al., 1998; Hellen & Sarnow, 2001). In addition to cross-linking experiments, hydroxyl radical footprinting studies were carried out with the use of tRNA analogs deprived of the 3'-terminal trinucleotide CCA; in these studies, conclusions on the 28S rRNA nucleotides participating in binding of tRNA CCA termini were made based on the comparison between the data with 28S rRNA isolated from the ribosomal complexes containing full size $tRNA^{Asp}$ and the results obtained with $tRNA^{Asp}$ lacking in the CCA end (Bulygin et al., 2012).

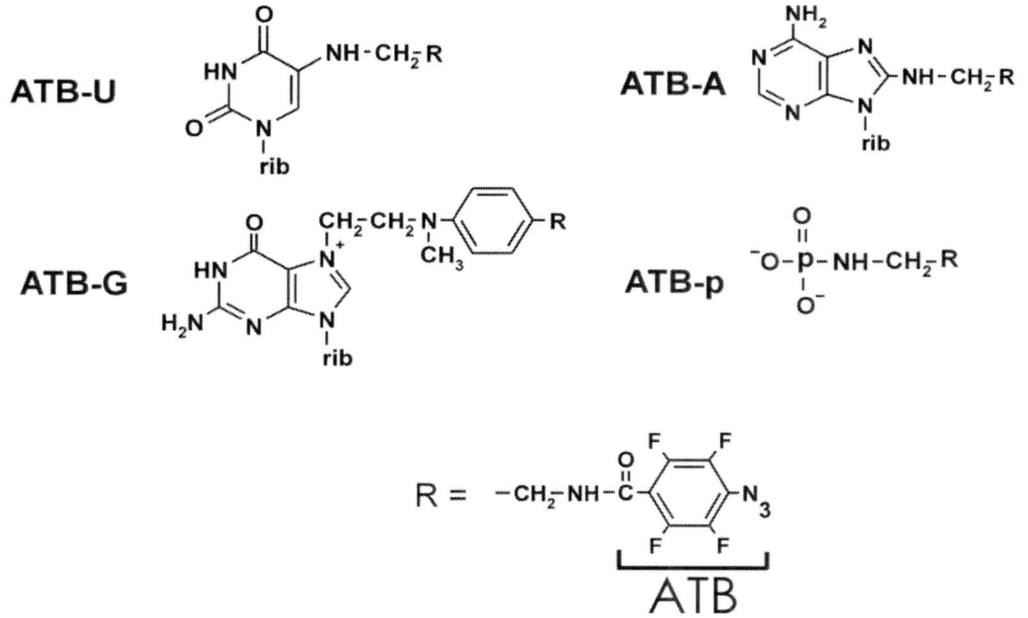

Figure 1. Structures of ATB-modified nucleosides and terminal phosphate in mRNA analogs.

mRNA Binding Center

18S rRNA Nucleotides and Ribosomal Proteins Neighboring mRNA

First systematic studies of the mRNA binding center of the human 80S ribosome were carried out with the application of a large set of short ATB-derivatized mRNA analogs

(Demeshkina et al., 2000; 2003a & b; 2005; Bulygin et al., 1997; 2002; 2003; Graifer et al., 2004; Molotkov et al., 2005; 2006) (Figure 4).

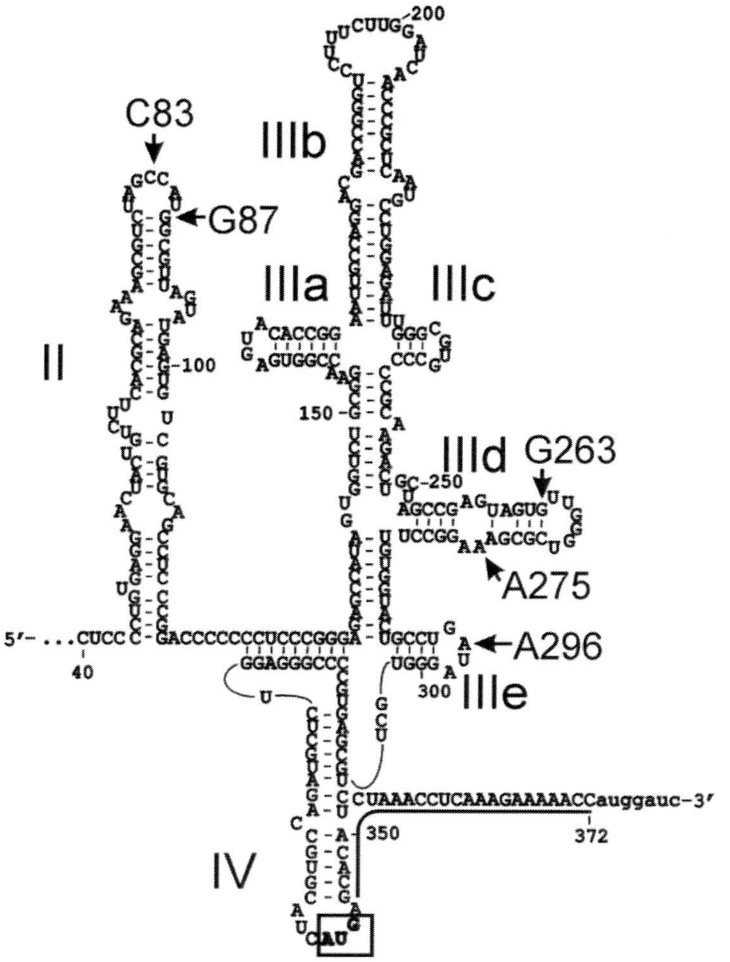

Figure 2. Secondary structure of the RNA transcript corresponding to HCV IRES according to Zhao & Wimmer, 2001. Domains and subdomains are indicated, start AUG codon is boxed, a part of the open reading frame (ORF) is underlined; nucleotides, which present in the transcript but are not the HCV IRES constituents, are given in lowercase letters.

These studies provided detailed information on the molecular environment of mRNA nucleotides in positions from -9 to +12 with respect to the first nucleotide of the P site codon on the 80S ribosome on the level of 18S rRNA nucleotides and identities of ribosomal proteins. Analysis of the data obtained made it possible to find for the first time universal features of the mRNA binding site in all kingdoms and to identify peculiarities specific to eukaryotic (or even to mammalian) ribosomes (Graifer et al., 2004; Molotkov et al., 2005; 2006). All cross-linked 18S rRNA nucleotides fell into the conserved core of the small subunit rRNA secondary structure, and positions of the cross-linked 18S nucleotides almost completely coincided with those of 16S rRNA nucleotides that interact or neighbor the mRNA nucleotides in the respective positions (Ogle et al., 2001; Yusupova et al., 2001; Selmer et al., 2006). These results experimentally confirmed the widely accepted idea on the conserved rRNA "core" of the ribosome. Data on mutual positioning of mRNA and 18S rRNA in the human ribosome obtained then with s^4U-containing relatively long mRNA analogs (Bulygin et al., 2005; 2009) were in a good agreement with the above mentioned results. In some cases, 18S rRNA nucleotides cross-linked to s^4U residues and to ATB-derivatized nucleotides at the same mRNA positions did not exactly coincide, but all these nucleotides were located very close (within 10-15 Å) to each other on the available 40S subunit spatial structures (Ben-Shem et al., 2011; Rabl et al., 2011). Notably, the cross-linking data showed that the molecular environment of the A site codon practically does not depend on the codon type (sense or stop codon) and on the presence of termination factor eRF1 (Bulygin et al., 2003). Data on 18S rRNA nucleotides contacting mRNA in rabbit 48S and 80S initiation complexes assembled with the use of a set of the appropriate initiation factors and mRNAs containing s^4U at positions from -26 to +11 (relatively to the first nucleotide of the AUG codon) (Pisarev et al., 2006; 2008) turned out to be very similar to those discussed above. All this shows that the conserved core of mammalian 18S rRNA accommodates mRNA to the ribosome very similarly during all steps of translation.

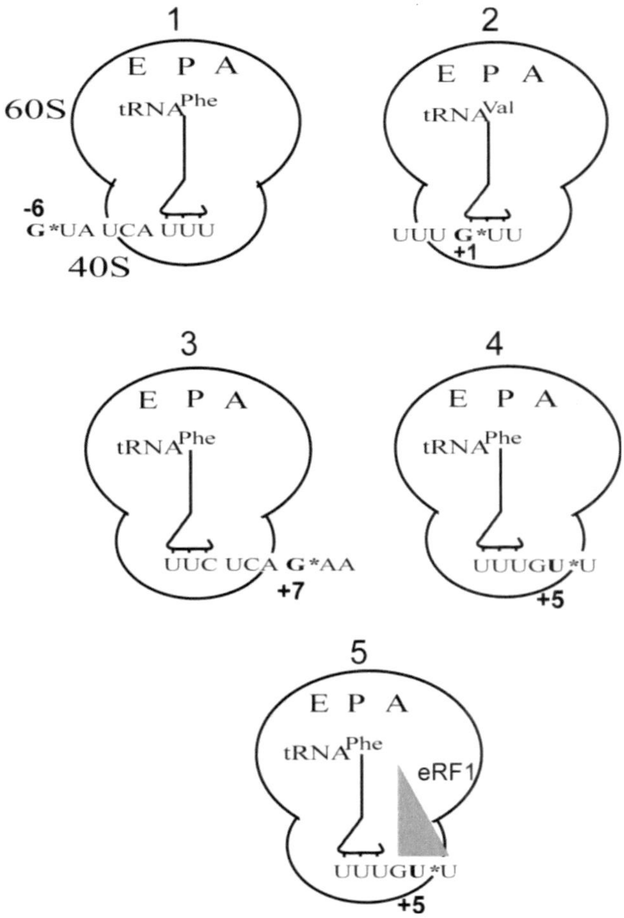

Figure 3. Examples of model 80S ribosomal complexes with ATB-modified mRNA analogs. A, P, and E are the tRNA binding sites. The derivatized nucleotides of mRNA analogs are in bold and marked with asterisks; their positions with respect to the first nucleotide of the P site codon are indicated.

The general representatives of the protein environment of mRNA on the ribosome are ribosomal proteins (rps) S26e and S14e neighboring mRNA part 5' of the E site codon (region of mRNA exit site), rpS15e contacting the A site codon (the decoding site), and rpS3e together with rpS2e, which are located close to mRNA part 3' of the A site codon (region of mRNA entry site) (Figure 5). Besides, with s^4U-containing mRNA analogs it was found that rpS5e contacts positions -3 and -4, and rpS28 interacts with position -6 (Figure 5) (Pisarev et al., 2008). One can see that the protein environment of mRNA on the 80S ribosome has features that have no homology in bacteria. So, rpS26e and rpS28e have no bacterial counterparts; rpS15e is homologous to rpS19p, but the latter is located away from the mRNA binding center in the 30S subunit (Yusupova et al., 2001; 2006). Similarities between the 40S and the 30S mRNA binding centers probably concern mammalian ribosomal proteins S14e, S3e and S2e and their prokaryotic homologs S11p, S3p and S5p that contact the same mRNA positions in the 40S and 30S subunits, which one can see from the comparison of X-ray data on the bacterial ribosome (Yusupova et al., 2001; 2006) with the reviewed cross-linking data on the mammalian ribosome.

Peptides of Ribosomal Proteins Involved in Formation of the mRNA Binding Center

Information on particular ribosomal protein peptides neighboring mRNA has been gained with three ribosomal proteins, rpS3e, rpS15e and rpS26e by using of ATB-derivatized mRNA analogs. Peptides cross-linked to labeled mRNA analogs were identified by means of a methodology based on the application of various proteolytic agents that cleave the polypeptide chain of the cross-linked protein at specific sites with subsequent separation of the resulting labeled modified peptides by polyacrylamide gel electrophoresis.

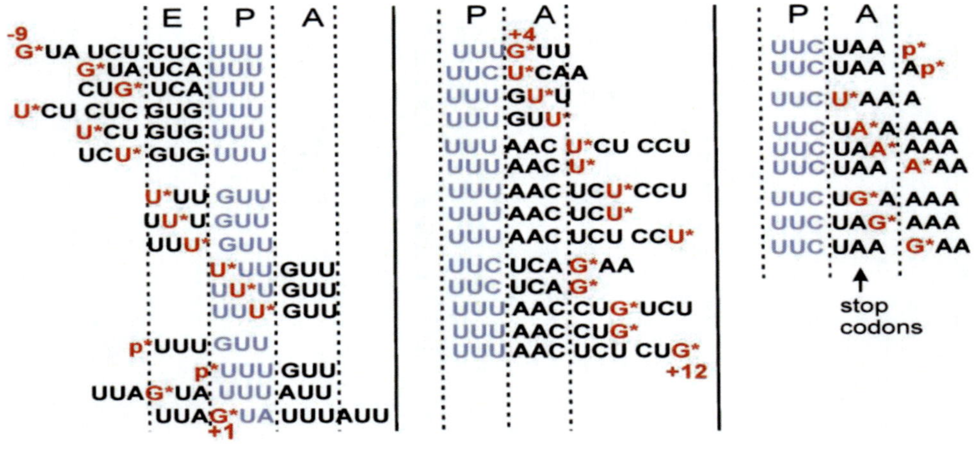

Figure 4. ATB-derivatized mRNA analogs applied for studying human translational machinery. The derivatized nucleotides or phosphates are in red and marked with asterisks. mRNA codons targeted to the ribosomal P site to fix the mRNA on the ribosome and thus to direct the derivatized nucleotide to the designed position are in blue. Position +1 conventionally corresponds to the first nucleotide of the P site codon; positions of mRNA nucleotides the upstream and downstream of this nucleotide are marked "-" and "+", respectively.

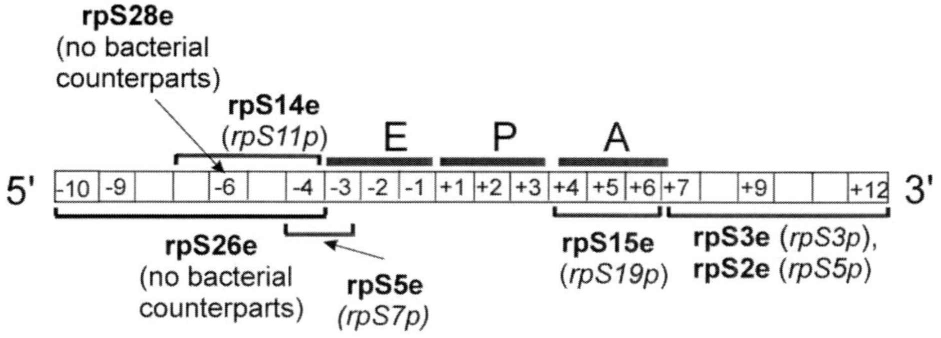

Figure 5. Ribosomal proteins that neighbor or contact particular positions of mRNA according to cross-linking data (Demeshkina et al., 2003b; Graifer et al., 2004; Molotkov et al., 2006; Pisarev et al., 2008). mRNA is shown as diagram, positions of the mRNA nucleotides on the ribosome are indicated, codons interacting with tRNAs are marked, bacterial counterparts are shown in brackets.

Mapping of the mRNA analogs cross-linking sites on the rpS3e was performed using only one proteolytic agent, CNBr, therefore, it was rather imprecise. So, it was found that N-terminal fragment (aa 2-127) of the protein was the main target for cross-linking of derivatized U, A or G in position +7, and C-terminal fragment (aa 190-236) cross-linked presumably to the derivatized U in position +9 or phosphate on a nucleotide in position +7; both fragments of rpS3e cross-linked to derivatized U in position +12 (Khairulina et al., 2008). Contacts of mRNA with the N-terminal half of rpS3e probably reflect similar features of the structures of mammalian and bacterial mRNA binding centers. This suggestion came from the comparison of the cross-linking results with X-ray data on positioning of mRNA in the 30S subunit. The X-ray data showed that mRNA part 3' of the A site codon contacts aa residues Arg131, Arg132, Lys135 and Arg164 of the rpS3p (a bacterial counterpart of rpS3e) (Yusupova et al., 2001), three of which (Arg131, Arg132 and Lys135) are located in the conserved fragment of the protein located in its N-terminal part (Khairulina et al., 2008). On the other hand, C-terminal part of rpS3p is not involved in the formation of the 30S ribosomal mRNA binding centre (Yusupova et al., 2001; 2006), therefore, cross-linking of mRNA analogs to the C-terminal fragment of rpS3e reflects a feature peculiar to mammalian ribosomes. X-ray crystallographic study of yeast 80S ribosome (Ben-Shem, et al., 2010; 2011) showed that arrangement of the ribosomal area corresponding to the region of mRNA entry site is substantially different in 70S and 80S ribosomes. In particular, dissimilarities have been found in protein-rRNA and protein-protein interactions involving rpS3e/rpS3p. Evidently, further more exact mapping of mRNA analogs cross-linking sites on rpS3e is required to study fine structure of the mRNA entry site on the 80S ribosome, which could make it possible to reveal actual extent of similarity in mRNA binding tracks in the 70S and 80S ribosomal entry site regions.

Contacts of mRNA codon with rpS15e at the ribosomal decoding site are inherent for the mammalian ribosome at all steps of translation, namely, at initiation (Pisarev et al., 2006; 2008), elongation and termination (Graifer et al., 2004; Khairulina et al., 2010). The contacts most probably involve nucleotides in the first and second positions of the A site codon since ATB-derivatized nucleotides in these positions cross-linked to rpS15e much more effectively than that in third position (Graifer et al., 2004), and s^4U cross-linked to rpS15e only from the first and second positions of the A site codon (Pisarev et al., 2008). Bacterial rpS19p, which is a homolog of rpS15e, is located mainly on the 30S subunit head, but its C-terminal tail extends towards the decoding site although does not reach it (Yusupova et al., 2001; Ogle et al., 2001). It was suggested that the tail of mammalian rpS15e reaches the decoding site in contrast to the tail of rpS19p (Graifer et al., 2004). This suggestion was experimentally confirmed in a study (Khairulina et al., 2010) reported that decapeptide 131-PGIGATHSSR-140 in the C-terminal fragment of rpS15e was the cross-linking target for the A site bound ATB-derivatized nucleotides of mRNA analogs. RpS15e has been mapped on the head of 40S subunit in several cryo-EM studies based on docking of known atomic structure of bacterial rpS19p into the 40S density map (Chandramouli et al., 2008; Taylor et al., 2009; Armache et al., 2010), but its unstructured C-terminal tail, which has almost no sequence homology with the tail of bacterial rpS19p (Khairulina et al., 2010), could not be mapped this way. Based on the cross-linking results and on similar locations of the structured parts of eukaryotic rpS15e and bacterial rpS19p in the small ribosomal subunits, a scheme of possible positioning of the C-terminal tail of human rpS15e with respect to the mRNA on the 40S ribosomal subunit has been created (Khairulina et al., 2010) (Figure 6). It is worth to mention that the main part of the unstructured C-terminal tail of rpS15e remains lacking in the 40S subunit atomic structures obtained by X-ray crystallography (Ben-Shem et al., 2010; 2011; Rabl et al., 2011). The last resolved amino acid residues in the C-terminal part of rpS15e are within 30 Å away from of the putative site for the A site codon of mRNA (deduced on the basis of similar positioning of the A site codon with respect to the conserved rRNA decoding site in prokaryotic and eukaryotic ribosomes).

The finding that a rpS15e C-terminal decapeptide takes part in the formation of the 40S ribosomal decoding site showed principal difference in the structural organizations of 40S and 30S decoding sites. So, in bacteria the ribosomal binding pocket for the A site mRNA codon is formed primarily by the 16S rRNA, which is responsible for all interactions of the ribosome with the codon and anticodon bases (reviewed in (Ramakrishnan, 2002; Noller, 2006; Steitz, 2008)). It is not yet clear how the C-terminal decapeptide rpS15e could participate in the translation process, and several hypotheses concerning its possible role have been suggested (Khairulina et al., 2010). Remarkably, this decapeptide was not found in the Protein data bank in any other protein than rpS15e; only short fragments of this sequence (not longer than penta- or hexapeptide motifs) were traced in several protein families. This points to an important role of a part of the unstructured C-terminal tail of rpS15e in the translation process in eukaryotes and probably in archaea. During initiation, as suggested earlier (Pisarev et al., 2008), it could be important for the initiation codon selection and/or for the stabilization of 48S complexes since rpS15e contacts mRNA nucleotides in positions +4 and +5 in these complexes. During elongation, it possibly contributes to the fixation of mRNA codon at the decoding site that, in turn, could provide higher accuracy of translation and less probability of a frameshift. During termination, the decapeptide possibly interacts with polypeptide chain release factor eRF1 and thus mediates stop codon decoding. This assumption is based on the data on shielding of rpS15e from contacts with mRNA by eRF1 (Chavatte et al., 2003). This factor has little sequence homology with the bacterial counterparts but is homologous to archaeal aRF1 (reviewed in Kisselev et al., 2003; Korostelev, 2011), therefore, interaction of C-terminal decapeptide of rpS15e with eRF1 could be its eukaryote/archae-specific function. This suggestion seems more attractive and grounded than others.

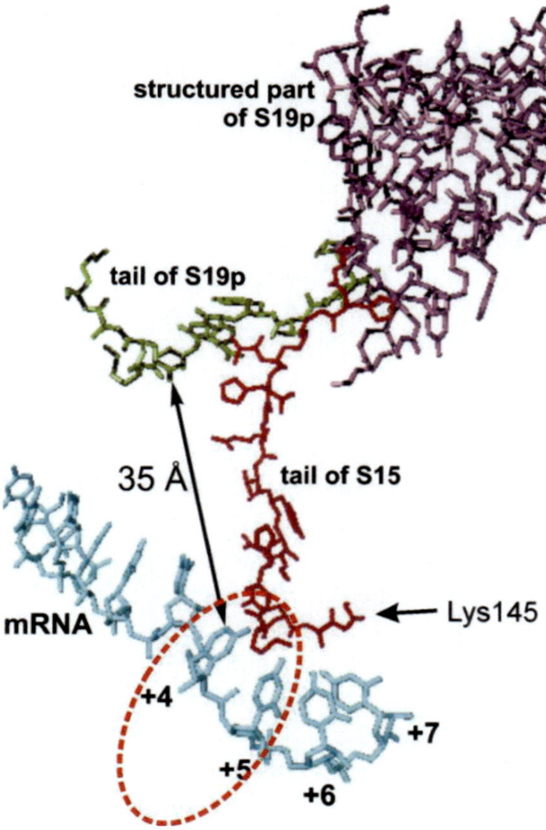

Figure 6. A scheme of possible positioning of the C-terminal sequence 120-145 of human rpS15e with respect to the mRNA on the 40S ribosomal subunit based on the cross-linking results and on the atomic model of the 70S ribosomal complex with mRNA and tRNAs (PDB ID code 1IBM, Ogle et al., 2001). All components of the complex except rpS19p and the mRNA were removed from the model (five last C-terminal amino acids of S19p in positions 89-93 in the atomic model are missing). The C-terminal sequence 120-145 of human rpS15e (presented in dark red) was built arbitrarily by sequential attachment of amino acid residues taken from the model 1IBM, and then attached virtually to the structured part of rpS19p (presented in purple) to fit the cross-linking results. Dotted circle covers first and second nucleotides of the A site mRNA codon and aa 137-140 in the tail of rpS15e that come the most closely to these nucleotides and therefore can contact them.

A single rpS26e fragment was found to be involved in the organization of the 80S ribosomal mRNA binding centre. So, cross-linking site of ATB-derivatized nucleotides of mRNA analogs in positions -6 and -9 was mapped to the fragment 60-71 of the protein (Sharifulin et al., 2012). Alignment of amino acid sequences of eukaryotic (from lower fungal to mammalian) and archaeal proteins from the rpS26e family showed that this fragment contains motif 62-YxxPKxYxK-70 conserved in eukaryotic but not in archaeal rpS26e. Analysis of the X-ray structure of the *Tetrahymena thermophila* 40S subunit (Rabl et al., 2011) showed that this motif is not implicated in the intra-ribosomal interactions, which led to a suggestion on its involvement in the translation process in a eukaryote-specific manner, e.g., in binding of eukaryote-specific ligands (Figure 7). One of these ligands could be eukaryotic initiation factor eIF3 that is involved in recruitment of the mRNA to the 40S subunit and has no counterparts in eubacteria and archaea (for a review, see (Rodnina & Wintermeyer, 2009)). This suggestion was based on the data on overlapping locations of rpS26e (Rabl et al., 2011) and eIF3 (Siridechadilok et al., 2005) on the 40S subunit (Figure 7) and on the data on contacts of mRNA nucleotides in position -8 and upstream of it with eIF3 (Pisarev et al., 2008). So, it was hypothesized that motif 62-YxxPKxYxK-70 of rpS26e is involved in binding of both the 5'-untranslated region of the mRNA and eIF3, and that this interaction underlies a mechanism of mRNA recruitment by eIF3 (Sharifulin et al., 2012). The role of the mentioned rpS26e motif could extend beyond the initiation step by its possible contribution to the maintenance of the mRNA path from the region of codon-anticodon interactions to the exit site during translation. Finally, the YxxPKxYxK motif could be a potential target for the action of some kind of regulatory factors modulating ribosomal function during translation since it is partly accessible at the surface of the 40S subunit solvent side (Rabl et al., 2011).

Although rpS26e lacks bacterial counterparts, in the 40S subunit this protein occupies a site homologous to that for rpS18p in the 30S subunit (Armache et al., 2010; Rabl et al., 2011) and both proteins neighbor the mRNA region 5' of the E-site codon in the small ribosomal subunit (Yusupova et al., 2001; Graifer et al., 2004; Pisarev et al., 2008). However, analysis of the X-ray structures of the 40S and 30S subunits showed that positioning of rpS18p with respect to 16S rRNA has almost nothing in common with positioning of rpS26e with respect to the 18S rRNA (Sharifulin et al., 2012) indicating for different roles of rpS18p and rpS26e in translation.

Notably, K66 from the 62-YxxPKxYxK-70 motif of rpS26e is acetylated in human cells (Choudhary et al., 2009). Acetylation is a reversible post-translational modification of proteins playing a key role in regulating gene expression; lysine acetylation preferentially regulates the functioning of large macromolecular complexes involved in diverse cell processes, such as modulation of chromatin architecture, cell cycle, splicing, nuclear transport, etc. (e.g., see (Choudhary et al., 2009) and refs therein). However, rpS26e isolated from rat (Louie et al., 1996) and human (Malygin & Karpova, 2010) 40S ribosomal subunits lack any post-translational modifications, so it was suggested that rpS26e is deacetylated before or during ribosome assembly. Since acetylated lysines lack positive charges and are therefore unable to interact with RNA phosphates, acetylation of rpS26e at K66 can deprive the protein of its ability to interact with mRNA. Therefore, deacetylation of the protein involved in ribosome assembly could be of major importance to ensure interaction of its 62-YxxPKxYxK-70 motif with mRNA. Further studies are required to clarify the role of 62-YxxPKxYxK-70 motif of rpS26e and acetylation of its K66 in translation.

Eukaryote-Specific Role of 2'OH Groups for the Binding of mRNA Codons at the 40S Ribosomal P Site

To complete the review on the features of mRNA binding center of the mammalian ribosome, it is worth to mention recent data obtained by studying cross-linking of mRNA and mDNA analogs to ribosomal proteins in human 80S ribosomal complexes (Graifer et al., 2012). It was found that both mRNA and mDNA analogs could cross-link to rpS3e and rpS2e at the mRNA entry site independently on the presence of tRNA, but only mRNA analogs were capable of a tRNA-dependent binding to human ribosomes and cross-linking to rpS26e in the mRNA binding centre. The conclusion was made that tRNA-independent binding of mRNA at the 80S ribosomal mRNA entry site does not involve 2'-OH groups of mRNA, but these groups are crucial for codon-anticodon interactions at the P site, which is in sharp contrast with the data on bacterial ribosomes showed that 30S/70S ribosomes do not discriminate single stranded DNA and RNA as templates for codon-dependent binding of tRNA at the P site (see Potapov et al., 1995, and refs therein). The data obtained with 80S ribosomes indicated the existence of mRNA-ribosome contacts that do not occur in bacteria. Analysis of the available X-ray data on the structures of 30S and 40S ribosomes revealed the only structural element of the small ribosomal subunit whose location with respect to the ribose-phosphate backbone of P site mRNA codon is different in 40S and 30S subunits. In the 40S subunit this element is located much closer to the backbone than in the 30S one. It was 18S rRNA fragment at the base of helix 44 in positions 1495-1499 (*E. coli* 16S rRNA numbering), and this fragment was suggested to be able to form eukaryote-specific contacts with mRNA codon at the P site that involve most probably 2'-OH of the codon first nucleotide and the phosphate between nucleotides 1497 and 1498. The mentioned 18S fragment belongs to the conserved core of the small subunit rRNA secondary structure, and presented above data give the first indication for dissimilarities in the interactions of mRNA codons with the core at the region of codon-anticodon interactions in 30S and 40S subunits.

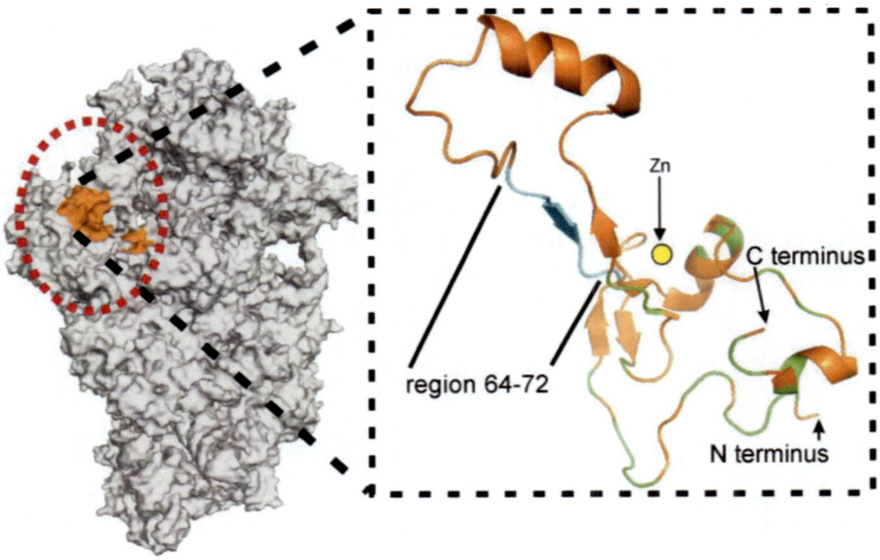

Figure 7. Positioning of rpS26e on the *T. thermophila* 40S subunit (PDB accession number 2xzn) (Rabl et al., 2011). General view (on the left) and zoomed structure of rpS26e extracted from the model (on the right). Sequence 64–72 corresponding to the conserved motif YxxPKxYxK is cyan, protein fragments interacting with the 18S rRNA are marked as green, location of the Zinc atom (presented as yellow circle) is shown. Dotted circle roughly shows the location of eIF3 on the 40S subunit according to Siridechadilok et al., 2005.

Stop Codon Recognition Site on eRF1

The structure of human eRF1, which has been deciphered by X-ray crystallography (Song et al., 2000), has almost no similarities with structures of bacterial RF1/RF2; sequences of the bacterial and eukaryotic termination factors are in major not homologous as well (see above). Y-shaped crystal structure of eRF1 consists of the N-terminal (N), middle (M) and C-terminal (C) domains (Figure 8) whose functional roles are well known from numerous biochemical *in vitro* studies, genetic *in vivo* studies, and bioinformatics studies (reviewed in Kisselev et al., 2003; Cheng et al., 2009). The conventional point of view is that the eRF1 structure is a tRNA-mimic. The C domain is responsible for the interaction with eRF3, the M domain mimics the tRNA acceptor stem and contains a universal GGQ motif common to all class-1 RFs. This motif is located at the tip of the M domain of eRF1 and is essential for peptidyl-tRNA hydrolysis at the ribosomal peptidyl transferase center. The N domain mimics tRNA anticodon arm and contains highly conserved YxCxxxF (positions 125-131) and NIKS (positions 61-64) motifs that play a critical role in stop codon recognition (reviewed in Kisselev et al., 2003). For all that, X-ray structure of human eRF1 does not fit ribosomal cavity for the A site tRNA since the distance between the GGQ and NIKS or YxCxxxF motifs in eRF1 is about 100 Å that is much larger than the distance between the acceptor end and the anticodon loop in a tRNA (typically, 75 to 80 Å). This implies that eRF1 can be adopted at the ribosomal A-site only when it undergoes major conformational rearrangements. Such rearrangements have been observed in the X-ray crystallographic studies of eRF1 complexed with eRF3 without ribosomes (Cheng et al., 2009) and in very recent cryo-EM study of the 80S ribosomal termination complex (Taylor et al., 2012). Positioning of mRNA stop signal with respect to eRF1 in the 80S ribosomal termination complexes was studied in detail by site-directed cross-linking with application of mRNA analogs and model complexes described above (Figure 3, complex 5). Data on cross-linking of mRNA analogs (Figure 4, the very right column) containing stop signal targeted to the A site to particular eRF1 peptides in 80S termination complexes are summarized in Table 1. It is seen that the stop signal recognition site of eRF1 is composed of three distinct parts proximal to the stop signal and to each other within the ribosomal A-site (Figure 8).

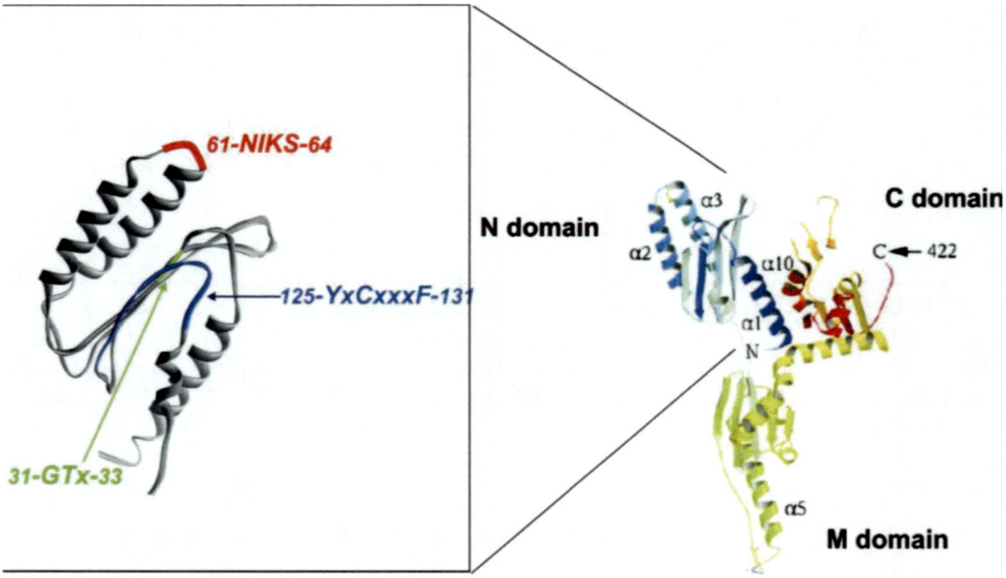

Figure 8. Spatial structure of human translation termination factor eRF1 according to Song et al., 2000 (right) and a magnified view of its N-domain (left) where conserved amino acid motifs are shown that were found to cross-link to derivatized nucleotides in the A site bound stop codon of mRNA analogs within the ribosomal termination complexes (Chavatte et al., 2002; Bulygin et al., 2010; 2011).

Arrangement of invariant U in the first stop codon position in the termination complexes was studied with application of two types of cross-linker, s^4U (Chavatte et al., 2002) and ATB-derivatized U (Bulygin et al., 2010), whose targets in the eRF1 were close to each other although not coincided (Table 1), most probably due to the different natures of the cross-linkers. Taken these results together, one can consider that pentapeptide IKSRV (positions 62-66) of human eRF1 is the closest neighbor of the universal stop codon uridine. From structural considerations it has been suggested that to distinguish U from C at least two amino acids are required (Saenger, 1984). Therefore, probably two or three residues of this peptide may be involved in recognition of the first U. Positioning of stop signal purines with respect to eRF1 was found to be drastically dependent on the their nature, which was displayed in different targets of cross-linking of derivatized A and G in the same stop signal positions (Bulygin et al., 2010; 2011) (Table 1). The application of the cross-linking results to the structure of eRF1 bound to the ribosome obtained by molecular modeling showed that peptide neighborhoods of guanines and adenines of stop signals are compatible with two alternative conformations of the N domain of eRF1 in the 80S termination complex (Bulygin et al., 2011). The key difference between these conformations of the N domain concerned the positioning of the stop signal purines with respect to the universally conserved dipeptide GT (positions 31-32) in eRF1 that neighbors or even interacts with guanines but is oriented more distantly from adenines. The capability of the ribosome-bound eRF1 to easily adopt alternative conformations to recognize adenines and guanines was suggested to be essential for recognition of all stop signals by the same class 1 termination factor in eukaryotes. The suggestion on conformational rearrangements of the N domain of eRF1 providing stop codon recognition has been developed in the NMR study of model fragments of the termination complex consisting of wild type or mutated N-domain of eRF1 and RNA oligonucleotides mimicking helix 44 of 18S rRNA, which contains the decoding site (Wong et al., 2012).

Table 1. Cross-linking sites of mRNA analogs on human eRF1. Note that the conserved NIKS and YxCxxxF motifs are in positions 61-64 and 125-131 of eRF1, respectively

Derivatized nucleotide		Cross-linking site	
Position in the stop signal*	Cross-linker and its site of attachment	**Major**	Minor
1 (Chavatte et al., 2002)	s^4U	**K63**	
1 (Bulygin et al., 2010)	ATB-U	**V66**	
2 (Bulygin et al., 2010)	ATB-G	**31-GT-32**	121-124
3 (Bulygin et al., 2010)	ATB-G	**31-GT-32**	F131
4 (Bulygin et al., 2010)	ATB-G	**31-GT-32**	L124, F131
2 (Bulygin et al., 2011)	ATB-A	**127-132**	
3 (Bulygin et al., 2011)	ATB-A	**121-132**	
4 (Bulygin et al., 2011)	ATB-A	**26-28**	121-124
3 (Bulygin et al., 2011)	ATB-3'-p	**26-28**	
4 (Bulygin et al., 2011)	ATB-3'-p	**26-28**	67-73

* Tetranucleotide consisting of a stop codon and the adjacent downstream purine is considered as stop signal.

Results on cross-linking of mRNA analogs derivatized at the 3'-phosphates to tripeptide 26-AAR-28 of eRF1 (Table 1) give an idea on positioning of the stop signal ribose-phosphate backbone with respect to eRF1 in the 80S termination complex. The mentioned tripeptide is not conserved, but in majority of organisms the eRF1 position 28 (human eRF1 numbering) is occupied by arginine or lysine (Liang et al., 2005). It has been reasonably suggested that the positively charged side chain of arginine/lysine in eRF1 position 28 interacts with the negatively charged phosphates 3' of adenosines in the 3rd and 4th positions of the termination tetraplet (Bulygin et al., 2011); this interaction cannot mediate discrimination of stop codon purines, but it may contribute to the energy required for the interaction of eRF1 with a stop codon.

HCV IRES Binding Site on the 40S Ribosome

Formation of the stable binary complex of HCV IRES with the 40S subunit where the initiation codon is positioned near the P site without any initiation factors plays the key role in the initiation of HCV gRNA translation (reviewed in Hellen & Sarnow, 2001; Otto & Puglisi, 2004; Hellen, 2009; Shatsky et al., 2010). However, only lesser part of numerous studies on HCV IRES concerns structural organization of its binding site on the 40S ribosomal subunit. Structural elements of HCV IRES that are crucially important for the binary complex formation are basal portion of domain III (mainly hairpins IIId and IIIe) and the four-way junction of subdomains IIIa, IIIb and IIIc (Figure 2) (Lukavsky, 2009). Cryo-EM data revealed that HCV IRES binds at the solvent side of the subunit in the region homologous to the mRNA exit site in the 30S subunit (Spahn et al., 2001). Two segments of the IRES extend from this exit channel, one (identified as upper part of domain II) toward the head, and the other (containing subdomains IIIb and IIIa/c) to the back of the platform (Figure 2). HCV-IRES-induced conformational change in the 40S region close to the mRNA entry channel was observed. In particular, the change involved opening of a latch between 18S rRNA helices 16 and 34; the authors assumed that this opening helps 3'-end of the coding part of the HCV RNA to thread into the mRNA entry channel (Spahn et al., 2001). IRES-driven initiation of HCV gRNA translation involves only few initiation factors including eIF3 (Otto & Puglisi, 2004; Locker et al., 2007; Shatsky et al., 2010). Cryo-electron microscopy reconstructions showed that eIF3 interacts with the HCV IRES bound to the 40S subunit and the 5'-cap binding complex eIF4F via the same domain (Siridechadilok et al., 2005). Detailed modeling of eIF3 and eIF4F onto the 40S ribosomal subunit enabled the authors to arrive to the conclusion that eIF3 uses HCV IRES or eIF4F in structurally similar ways to position the mRNA strand near the exit site of the 40S subunit, promoting initiation complex assembly.

The mentioned reports did not provide particular information on ribosomal components forming the HCV IRES binding site on the 40S subunit. In more detailed cryo-EM study with the IRES complexed with the 80S ribosome (Boehringer et al., 2005), two ribosomal proteins, namely, rpS5e on the head of the subunit interface and rpS14e on the platform, were mapped close to the apex of domain II (Figure 2). However, domain II, in contrast to subdomains IIId and IIIe and the four-way junction IIIabc, does not contribute to the affinity of HCV IRES to the 40S subunit (Lukavsky, 2009). Thus, ribosomal components interacting with HCV IRES subdomains crucial for its binding to the 40S subunit remained unknown from the cryo-EM studies. These components have been revealed by a number of other approaches, primarily by site-directed cross-linking.

With the use of direct UV-induced cross-linking it was found that in the binary complex with the 40S subunit IRES cross-links with only one protein identified finally as rpS5e (Fikushi et al., 2001); in contrast, HCV IRES with uridines randomly substituted by 4-thiouridines cross-linked to a wide set of proteins excluding rpS5e (Otto et al., 2002) (Table 2). Other sets of cross-linked proteins were obtained with HCV IRES derivatives bearing an ATB-derivatized nucleotides in the selected locations (see Figure 2), which made it possible to reveal ribosomal environment of the particular structural elements of the HCV IRES (Laletina et al., 2006; Babaylova et al., 2009) (Table 2). Ex facte, all reported cross-linking data seem rather contradictory. In particular, the sets of cross-linked proteins found with the s^4U-containing and the ATB-derivatized HCV IRES overlap by the single rpS16e. The explanation of these differences could be that the proteins found in (Otto et al., 2002) cross-linked with those s^4U residues that were located in positions, which were not derivatized with ATB groups in (Laletina et al., 2006; Babaylova et al., 2009), and vice versa. This was actually evidenced in the recent report on protection of lysines of 40S proteins from chemical modification by the bound HCV IRES and its mutant forms deprived of selected subdomains, an approach that was called "protein footprinting" (Malygin et al., 2012). In the mentioned report it was found that full sized HCV IRES protects from chemical modification rpS27e, rpS10 and RACK1/rpS2e/rpS3a (these proteins were not resolved). Deletion of the initial part of the open reading frame (ORF) of the HCV gRNA resulted in the disappearance of the rpS10e protection, and the deletion of both domain II and the ORF led to a weakening of the RACK1/S2/S3a protection in addition to the loss of the rpS10e protection. Comparison of these results with known structural and biochemical data (Boehringer et al., 2005; Babaylova et al., 2009; Ben-Shem et al., 2011) made it possible to reveal structural elements of the IRES contacting the exposed lysines of the mentioned ribosomal proteins. Thus, it was found that exposed lysines of rpS27e are mainly involved in the binding of HCV IRES region formed by the junction of subdomains IIIa, IIIb and IIIc with the central stalk of domain III, and that several lysines of rpS10e participate in the binding of HCV IRES region corresponding to the initial part of ORF of the viral gRNA. In addition, it was concluded that lysine residues of rpS3a are involved in the binding of the IRES domains II and III. As for cross-linking of HCV IRES derivatives to rpS2e, rpS3e and rpS15e (Otto et al., 2002), it could occur due to s^4U residues substituted U in the IRES part corresponding to the coding sequence (after AUG codon) in domain IV because these proteins belong to the mRNA binding center (see above). Data on positioning of HCV IRES on the 40S subunit are presented in Figure 9.

It is worth mentioning that in none of studies cross-links of HCV IRES to the 18S rRNA were reported. However, the lack of cross-links does not necessarily imply the absence of the contacts since there could not be a suitable target for cross-linking, or the photoactivated group could be oriented unfavorably with respect to the rRNA to obtain a cross-link. Thus, HCV IRES-rRNA contacts are to be further investigated.

In general, cross-linking results obtained with ATB-derivatized HCV IRES (Table 2) are in a good agreement with location of HCV IRES on the 40S subunit determined by cryo-EM. The only pronounced dissimilarity is that, according to the cross-linking data, rpS5e neighbors subdomain IIIe rather than domain II (Spahn et al., 2001). This in fact is not a real discrepancy because the absence of cross-link cannot indicate lack of neighborhood or even interaction of the IRES with rpS5e (see above). Proteins rpS16e and rpS14e cross-linked instead of rpS5e to the IRES derivatized at the subdomain IIb (Table 2), which is not unexpected because all these proteins are located rather close to each other on the 40S subunit (Figure 9), and the maximum distance separating the derivatized RNA nucleotide and the reactive nitrogen of the cross-linker could be up to 14 Å (Graifer et al., 2004). The same reasons as with rpS5e may explain the lack of cross-linking of HCV IRES derivatives to rpS25e, whose specific requirement for translation initiation of HCV gRNA was revealed with the use of yeast genetics (Landry et al., 2009).

To date, it remains largely unknown whether proteins found in the cross-linking studies are actually involved in the HCV IRES binding to the 40S subunit. Such kind information has been gained so far only concerning rpSA (p40). This protein is not a constant ribosomal component, and it is bound to 40S subunits in the polysomes during active cell growth while dissociation of polysomes leads to a partial loss of rpSA from the 40S subunits (Garcia-Hernandez et al., 1996). In accordance with these, it was found that 40S subunits isolated from human placenta vary in the extent of the rpSA loss, and subunits lacking in this protein can directly bind recombinant human rpSA (Kossinova et al., 2008). Such mobility of rpSA points to its loose contacts with the subunit. Recently it was reported that vertebral-specific C-terminal domain (CTD) of rpSA (fragment 236-262) mainly contributes to the binding of the protein to the 40S subunit (Kossinova et al., 2008; Malygin et al., 2011). According to hydroxyl radical footprinting data, the binding partner of this rpSA fragment in the 40S subunit is helix 44 of the 18S rRNA (Malygin et al., 2011), while the conserved core of the protein contacts only ribosomal proteins (Ben-Shem et al., 2011; Rabl et al., 2011). Remarkably, affinity of HCV IRES to 40S subunits strongly depended on their rpSA content (Malygin et al., 2009). Moreover, monoclonal antibodies against CTD of rpSA, which significantly contributes to the protein affinity to the 40S subunits, blocked HCV IRES binding to the subunits, but had no effect on binding of Phe-tRNA to the poly(U)-programmed ribosomes (Malygin et al., 2009). It turned out that the antibodies affect also translation of various mRNAs in a cell-free system. It was found that binding of the antibodies to their epitops in 40S subunits inhibits mRNA translation independently on the nature of its 5'-UTR (5'-UTR of actin mRNA or HCV IRES), but the effect is larger with the IRES-containing mRNA (Malygin et al., 2009). Since the antibodies were against CTD of rpSA, one can suggest that this protein domain is a key player in accommodation of HCV IRES to the 40S subunit. The results obtained with mRNA containing 5'-UTR of actin mRNA suggested that rpSA has a regulatory function in the process of translation of cellular mRNAs. The protein, when bound to the 40S subunit, could enhance or coordinate its interaction with initiation factor eIF3, whose binding site is close to the site of the conserved part of rpSA (see Siridechadilok et al., 2005; Ben-Shem et al., 2011; Rabl et al., 2011), and thus activate translation initiation. On the contrary, dissociation of rpSA could lead to decrease of the eIF3 affinity to 40S subunits, resulting in the reduction of the total translation efficacy. This implies that the conserved part of rpSA on the 40S subunit is involved in the interaction with the factor that, in turn, is likely implicated in binding with rpS26e (see above). All this is in a good accordance with close locations of rpS26e and the conserved part of rpSA on the 40S subunit (Rabl et al., 2011).

Region of the Peptidyl Transferase Center

The function of the ribosomal peptidyl transferase center (PTC) is the same in all kingdoms, and the secondary structure of the 23S/28S rRNA fragment forming the PTC is strongly conserved, so one could expect that positioning of tRNA CCA-ends are the same in 50S and 60S ribosomal subunits. However, the first cross-linking study with human 80S ribosomes carried out with tRNA analog bearing s^4U as an additional nucleotide attached to the 3'-terminal A revealed pronounced differences in the environments of the tRNA 3'-terminus in prokaryotic and eukaryotic ribosomes together with the expected similarities (Bulygin et al., 2008). These differences were displayed in dissimilarities between the set of the cross-linked nucleotides of the 28S rRNA and the set of candidate targets predicted by molecular modeling. It was suggested that the dissimilarities relate to different dynamic properties of ribosomes from various kingdoms. Indeed, different dynamic properties of mammalian and prokaryotic ribosomal complexes have been reported with application of cryo-electron microscopy and single-molecule FRET (Budkevich et al., 2011).

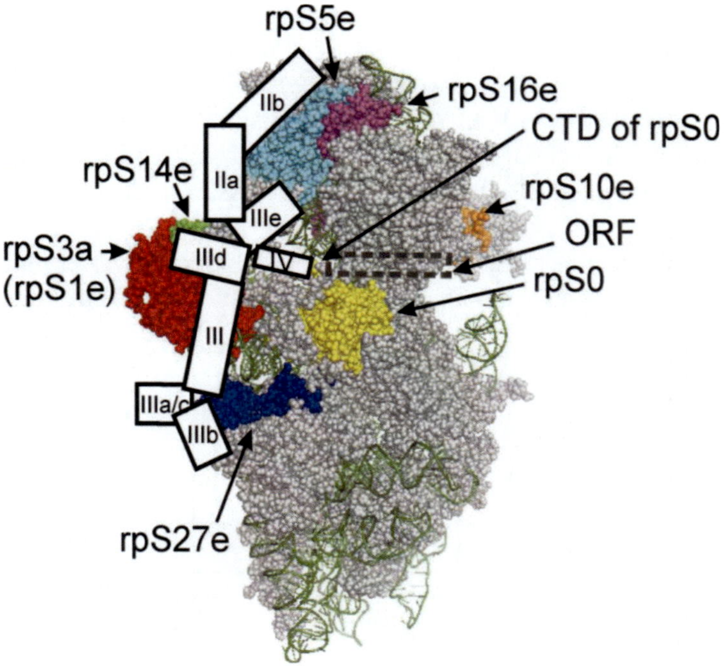

Figure 9. Application of site-directed cross-linking and protein footprinting data on the positioning of HCV IRES on the human 40S ribosomal subunit (Otto et al., 2002; Laletina et al., 2006; Babaylova et al., 2009; Malygin et al., 2012) to the atomic model of yeast 40S subunit (Ben-Shem et al., 2011; PDB entries 3U5B and 3U5C, view from the solvent side). IRES subdomains are shown as white rectangles; position of the subdomain IIb is shown according to cryo-EM data (Spahn et al., 2001). Ribosomal proteins are represented as surface, proteins belonging to the HCV IRES binding site are marked as follows: S0 (a counterpart of higher eukaryotic rpSA/p40), yellow; S3a (S1e), red; S5, cyan; S10, orange; S14, green; S16, magenta; S27, blue, other proteins are in grey; contour of the 18S rRNA is green. Dashed rectangle corresponds to the part of open reading frame (ORF) present in the HCV IRES, which is located on the subunit interface (on the back side of the presented view). Proteins contacting the ORF are not shown.

A comprehensive study of particular nucleotides of 28S rRNA surrounding CCA termini of tRNAs in the human 80S ribosome has been performed with the use of site-directed cross-linking s^4U-containing tRNA analogs together with hydroxyl radical footprinting (Bulygin et al., 2012). The rRNA nucleotides whose bases make intimate contacts with penultimate or 3'-terminal nucleotides of tRNA bound at the A or the P site of human 80S ribosomes were identified by the cross-linking with tRNA analogs bearing s^4U in position 75 or 76, and hydroxyl radical footprinting revealed 28S rRNA nucleotides involved in accommodation of CCA ends to the PTC region of the ribosome. These approaches, when applied for studying bacterial PTC (Moazed & Noller, 1989; Wower et al., 2000; Bocchetta et al., 2001), provided reliable results that turned out to be in a good general agreement with later X-ray crystallography data. Application of the cross-linking results to the X-ray derived model of the archaeal 50S subunit where all tRNA contacts are known (Schmeing et al., 2003; Korostelev et al., 2006; Selmer et al., 2006) and to the recently reported X-ray derived model of the yeast 80S ribosome (Ben-Shem et al., 2011) showed that, in general, the ribosomal components corresponding to A, P and E sites are similar in the archaeal and yeast large subunits. For all that, the targets of cross-linking of the P site s^4U-substituted tRNA analogs were located in the 60S ribosomal regions corresponding to both the P site and the E site positions of the tRNA acceptor ends in the archaeal 50S subunit, which implied that in the 80S ribosome the P site tRNA was present at both P/P and P/E states (Figure 10). Similarly, the A site tRNA was present in both A/A and A/P states. Unexpectedly, hydroxyl radical footprinting, which reveals interactions implicating the ribose-phosphate backbone, detected only P/P and A/P states with the P site and the A site tRNAs, respectively (Figure 10).

Table 2. Ribosomal proteins cross-linked to HCV IRES derivatives in binary complexes of these derivatives with mammalian 40S ribosomal subunits

Derivatized nucleotides and their positions in HCV IRES	Location of derivatized nucleotide in the HCV subdomain	Cross-linked ribosomal proteins	Reference
ATB-G263	IIId	S3a, S14e, S16e	Babaylova et al., 2009
ATB-A275		S3a, S14e, S16e	Babaylova et al., 2009
ATB-A296	IIIe	S3a, p40 (SA), S5, S16	Laletina et al., 2006
ATB-C83	Apical loop of domain II	S14e, S16e	Babaylova et al., 2009
ATB-G87	Near the apical loop of domain II	Not detected	Laletina et al., 2006
s^4U	Scattered randomly over HCV IRES	S2, S3, S10, S15, S16, S18, S27	Otto et al., 2002
None	Unmodified IRES	S5e	Fukushi et al., 2001

The explanation why A/A and P/E states were not detected by footprinting was that these tRNA states are unstable and therefore could be trapped only by cross-linking. On the other hand, footprinting revealed a number of 28S nucleotides in the P site region that were not detected by site-directed cross-linking; these nucleotides were assumed to interact with ribose-phosphate backbone rather than with bases of tRNA acceptor ends. It is worth to mention that 28S rRNA nucleotides cross-linked to CCA-ends of tRNA derivatives or protected by them in major did not exactly correspond to the 23S rRNA nucleotides interacting with the CCA-ends in the 50S subunit. An unexpected finding in the discussed study was the lack of both cross-links and footprints in the region of the tip of helix 80 of the large rRNA whose nucleotides (2251 and 2252 in *E.coli* numbering) contact CCA-terminus of the P site tRNA in the 50S subunit according to X-ray data obtained in various groups (Steitz, 2005; Selmer et al., 2006; Korostelev et al., 2006; Simonovic & Steitz, 2008). This finding implies that helix 80 is involved in the binding of the P site CCA termini only to 50S subunits but not to 60S subunits. The detailed comparative analysis of the X-ray models of archaeal and yeast large ribosomal subunits revealed that orientations of 23S rRNA nucleotides G2251 and G2252 at the tip of H80 in the 50S subunit and of the respective large rRNA nucleotides in the 60S subunit are similar as well as contacts of these nucleotides with other ribosomal components. Thus, the lack of interactions of the P site tRNA acceptor end with H80 does not relate to different structural organization of H80 in the 60S and 50S subunits. Taken together, all found dissimilarities in the positioning of the tRNAs CCA-termini regarding the 28S rRNA reflect different dynamic properties of mammalian and prokaryotic ribosomal complexes (see above).

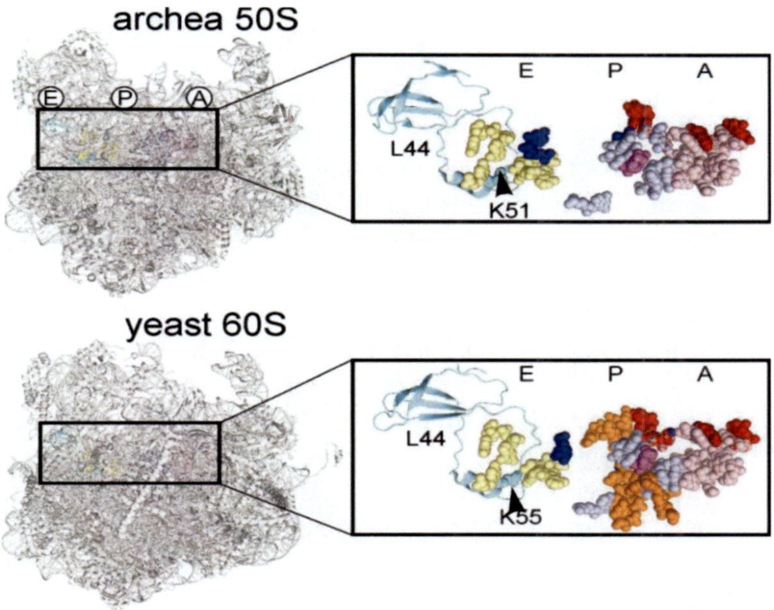

Figure 10. Application of the cross-linking and footprinting data on positioning of tRNA CCA-termini on the human ribosomes to the 3D structures of H.marismortui 50S subunits (Schmeing et al., 2003; PDB entry 1QVF) and of *S.cerevisiae* 60S subunits (Ben-Shem et al., 2011; PDB entries 3U5D and 3U5E). Nucleotides of the large rRNA located at A, P and E sites are colored in pink, light-violet and yellow, respectively. Cross-linking data are given in red and blue (with the A and the P site tRNA analogs, respectively), and the footprinting data are presented in orange. Nucleotides of large rRNA corresponding to G2251 and G2252 in the *E.coli* 23S rRNA are colored in magenta. Ribosomal protein L44e (homologous to human L36AL) is shown in light green. Positions of lysines corresponding to K53 in the human rpL36AL are indicated by arrows.

The evidence for the P/E state of the P site tRNA in the 80S ribosome has been obtained also when contacts of the 3'-terminus of the P site tRNA with ribosomal proteins were monitored with the use of 2',3'-dialdehyde derivatives of tRNA (Baouz et al., 2009). It was demonstrated that these contacts involve the only large ribosomal protein rpL36AL. This protein was found to contact all three nucleotides of the CCA-terminus via its Lys53, and the most effective cross-links were detected with the cytidines (Hountondji et al., 2012). rpL36AL belongs to the L44e protein family present in all eukaryotic and archaeal ribosomes; it is known that rpL44e in the archaeal 50S subunit contacts cytidines of the E site tRNA CCA terminus (Schmeing et al., 2003), which was the basis of the conclusion on the P/E state of the P site tRNA in the 80S ribosome. Since the 3'-oxidized tRNA analogs were unable to bind at the ribosomal E site, it was suggested that the modified tRNA might bind first to the classical P/P site before it flips spontaneously into the P/E site, where it is sampled by the cross-linking with Lys 53 of rpL36AL (Hountondji et al., 2012). The same suggestion concerns as well s^4U-substituted tRNA analogs since they were also unable to bind at the ribosomal E site (Bulygin et al., 2012). These results together with the discussed above cryo-EM data (Budkevich et al., 2011) point that CCA end of the P site tRNA oscillates between the classical P/P and the hybrid P/E positions. It was noticed that Lys53 of rpL36AL is located close to the 49-GGQ-51 tripeptide, which is conserved in all eukaryotic ribosomal proteins belonging to the L44e family, and is identical to the universally conserved motif of release factors implicated in promoting peptidyl-tRNA hydrolysis (Hountondji et al., 2012). One can suggest that functional role of rpL44e/rpL36AL relates to facilitation of translocation of deacylated tRNA after peptide bond formation from the P to the E site. High affinity of the CCA end of the deacylated P site tRNA to rpL44e/rpL36AL could be one of driving forces for the translocation.

In the conclusion, it is worth to mention recently reported data concerning hydroxylation of human rpL8e and rpL27a at definite His residues that occurs in the cell nucleolus (Ge et al., 2012). However, upon hypoxia the level of hydroxylation of these proteins strongly decreases because of inhibition of the respective oxygenase catalyzing the protein hydroxylation. On the atomic model of the yeast 80S ribosome (Ben-Shem et al., 2011) ribosomal proteins homologous to human rpL8 and rpL27a are located in the regions proximal to the sites where tRNA acceptor termini are bound when tRNA adopts the P/P and P/E states, respectively. One can suggest that the hydroxyl groups of these proteins are involved in either in the maintenance of functionally competent structure of the 28S rRNA in the PTC region or in the interaction with tRNA termini, and these interactions are distorted upon hypoxia leading to derange of the protein synthesis. A possibility of translational regulation by oxygenases was discussed (Ge et al., 2012). Further investigations are required to clarify these issues.

Conclusion

Although atomic structures of mammalian translational complexes were not so far obtained, significant knowledge is accumulated on the fine details of molecular environment of ribosomal ligands bound in these complexes from biochemical studies, carried out in major by site-directed cross-linking. Identification of 18S rRNA nucleotides and ribosomal proteins peptides taking part in the formation of the ribosomal mRNA binding and peptidyl transferase centers made it possible to evaluate extent of their similarity with the respective centers of prokaryotic ribosomes. As expected, the conserved features of the ribosomal functional sites are formed presumably by the rRNAs. So, the conserved core of small subunit rRNA almost identical in all kingdoms is involved in the structural organization of 80S ribosomal mRNA binding center and the core of domain V of the 28S rRNA forms the peptidyl transferase center. However, interactions of the conserved rRNA cores with RNA ligands in the human ribosome are not completely similar with those in the bacterial ribosome. In particular, some interactions occurring in the bacterial system, does not exist in human one, and vice versa. So, interaction of the P site tRNA acceptor end with H80 of the large rRNA inherent for the prokaryotic 50S subunit does not exist in human, and a contact of 18S rRNA with 2'-OH of the P site codon, which takes place in the human 40S subunit, does not occur in the bacterial 30S one. Stronger dissimilarities concern functional sites of the ribosome that are formed by ribosomal protein peptides, which turned out to have no homology in bacteria but are conserved in eukaryotes, and some of them also in archaea. However, our knowledge on the specific structural elements of the human ribosomal functional sites goes in advance of the understanding of the exact roles of these elements in the translation process requiring further special investigations.

Site-directed cross-linking and protein footprinting studies with HCV IRES derivatives made it possible to reveal 40S ribosomal components implicated in the formation of the HCV IRES binding site, a functional center that is specific to the mammalian ribosome. Involvement of ribosomal proteins in its formation implies participation of eukaryote-specific proteins or peptides in the HCV IRES binding. Indeed, almost a half of the set of ribosomal proteins found at the HCV IRES binding site have no bacterial counterparts, and rpSA is involved in the IRES binding by its vertebrate-specific C-terminal domain. These data shed light on the structural aspects of the HCV IRES interaction with the 40S subunit outlining the key role of ribosomal proteins in this interaction, but the mechanism by which the IRES manages to position start codon in the correct place on the 40S subunit is to be elucidated. The most comprehensive knowledge has been obtained concerning the stop codon recognition site on eRF1. Numerous genetic *in vivo* and bioinformatics studies revealed eRF1 peptides that are crucial for its polypeptide chain release activity and specificity towards particular stop codons, and the site-directed cross-linking studies revealed peptides that are directly involved in the formation of the stop codon recognition site. These studies provided new insights into molecular mechanisms of translation termination in eukaryotes and made it possible to get an idea how eRF1 discriminates stop codon purines.

References

Aitken, C. E. & Lorsch, J. R. (2012) A mechanistic overview of translation initiation in eukaryotes. *Nat. Struct. Mol. Biol.*, 19, 568-576.

Armache, J.-P., Jarasch, A., Anger, A. M., Villa, E., Becker, T., Bhushana, S., Jossinet, F. et al. (2010) Localization of eukaryote-specific ribosomal proteins in a 5.5-Å cryo-EM map of the 80S eukaryotic ribosome. *Proc. Natl. Acad. Sci. USA*, 107, 19754-19759.

Babaylova, E., Graifer, D., Malygin, A., Stahl, J., Shatsky, I. & Karpova, G. (2009) Positioning of subdomain IIId and apical loop of domain II of the hepatitis C IRES on the human 40S ribosome. *Nucleic Acids Res.*, 37, 1141–1151.

Ban, N., Nissen, P., Hansen, J., Moore, P. B. & Steitz, T. A. (2000) The complete atomic structure of the large ribosomal subunit at 2.4 Å resolution. *Science,* 289, 905–920.

Baouz, S., Woisard, A., Sinapah, S., le Caer, J.-P., Argentini, M., Bulygin, K., Aguié, G. & Hountondji, C. (2009) The human large subunit ribosomal protein L36A-like contacts the CCA end of P-site bound tRNA. *Biochimie*, 91, 1420-1425.

Ben-Shem, A., Jenner, L., Yusupova, G. & Yusupov, M. (2010) Crystal structure of the eukaryotic ribosome. *Science,* 330, 1203–1209.

Ben-Shem, A., Garreau de Loubresse, N., Melnikov, S., Jenner, L., Yusupova, G. & Yusupov, M. (2011) The structure of the eukaryotic ribosome at 3.0 Å resolution. *Science,* 334, 1524-1529.

Bocchetta, M., Xiong, L., Shah, S. & Mankin, A.S. (2001) Interactions between 23S rRNA and tRNA in the ribosomal E site. *RNA,* 7, 54-63.

Boehringer, D., Thermann, R., Ostareck-Lederer, A., Lewis, J. D. & Stark, H. (2005) Structure of the hepatitis C virus IRES bound to the human 80S ribosome: remodeling of the HCV IRES. *Structure,* 13, 1695-1706.

Budkevich, T., Giesebrecht, J., Altman, R. B., Munro, J. B., Mielke, T., Nierhaus, K. H., Blanchard, S. C. & Spahn, C. M. (2011) Structure and dynamics of the mammalian ribosomal pretranslocation complex. *Mol. Cell,* 44, 214-224.

Bulygin, K. N., Graifer, D. M., Repkova, M. N., Smolenskaya, I. A., Veniyaminova, A. G. & Karpova, G. G. (1997) Nucleotide G-1207 of 18S rRNA is an essential component of the human 80S ribosomal decoding center. *RNA,* 3, 1480-1485.

Bulygin, K. N., Repkova, M. N., Ven'yaminova, A. G., Graifer, D. M., Karpova, G. G., Frolova, L. Yu. & Kisselev, L. L. (2002) Positioning of the mRNA stop signal with respect to polypeptide chain release factors and ribosomal proteins in 80S ribosomes. *FEBS Lett.,* 514, 96-101.

Bulygin, K. N., Demeshkina, N. A., Frolova, L. Yu., Graifer, D. M., Ven'yaminova, A. G., Kisselev, L. L. & Karpova, G. G. (2003) The ribosomal A site-bound sense and stop codons are similarly positioned towards the A1823-A1824 dinucleotide of the 18S ribosomal RNA. *FEBS Lett.,* 548, 97-102.

Bulygin, K., Chavatte, L., Frolova, L., Karpova, G. & Favre, A. (2005) The first position of a codon placed in the A site of the human 80S ribosome contacts nucleotide C1696 of the 18S rRNA as well as proteins S2, S3, S3a, S30 and S15. *Biochemistry,* 44, 2153-2162.

Bulygin, K, Favre, A., Baouz-Drahy, S., Hountondji, C., Vorobjev, Yu., Ven'yaminova, A., Graifer, D. & Karpova, G. (2008) Arrangement of 3'-terminus of tRNA on the human ribosome as revealed from cross linking data. *Biochimie,* 90, 1624-1636.

Bulygin, K., Baouz-Drahy, S., Graifer, D., Favre, A. & Karpova, G. (2009) Sites of 18S rRNA contacting mRNA 3′ and 5′ of the P site codon in human ribosome: A cross-linking study with mRNAs carrying 4-thiouridines at specific positions. *Biochim. Biophys. Acta*, 1789, 167-174.

Bulygin, K. N., Khairulina, Yu. S., Kolosov, P. M., Ven'yaminova, A. G., Graifer, D. M., Vorobjev, Yu. N., Frolova, L. Yu., Kisselev, L. L. & Karpova, G. G. (2010) Three distinct peptides from the N domain of translation termination factor eRF1 surround stop codon in the ribosome. *RNA,* 16, 1902-1914.

Bulygin, K. N., Khairulina, Yu. S., Kolosov, P. M., Ven'yaminova, A. G., Graifer, D. M., Vorobjev, Yu. N., Frolova, L.Yu. & Karpova, G. G. (2011) Adenine and guanine recognition of stop codon is mediated by different N domain conformations of translation termination factor eRF1. *Nucleic Acids Res.,* 39, 7134–7146.

Bulygin, K., Malygin, A., Hountondji, C., Graifer, D. & Karpova, G. (2012) Positioning of CCA-arms of the A- and the P-tRNAs towards the 28S rRNA in the human ribosome. *Biochimie,* doi: 10.1016/j.biochi.2012.09.010.

Chandramouli, P., Topf, M., Menetret, J.-F., Eswar, N., Cannone, J. J., Gutell, R. R., Sali, A. & Akey, C. W. (2008) Structure of the mammalian 80S ribosome at 8.7 Å resolution. *Structure,* 16, 535–548.

Chavatte, L., Frolova, L., Kisselev, L. & Favre, A. (2001) The polypeptide chain release factor eRF1 specifically contacts the s(4)UGA stop codon located in the A site of eukaryotic ribosomes. *Eur. J. Biochem.,* 268, 2896-2904.

Chavatte, L., Seit-Nebi, A., Dubovaya, V. & Favre, A. (2002) The invariant uridine of stop codons contacts the conserved NIKSR loop of human eRF1 in the ribosome. *EMBO J.,* 21, 5302-5311.

Chavatte, L., Frolova, L., Laugâa, P., Kisselev, L. & Favre, A. (2003) Stop codons and UGG promote efficient binding of the polypeptide release factor eRF1 to the ribosomal A site. *J. Mol. Biol.*, 331, 745-758.

Cheng, Z., Saito, K., Pisarev, A. V. et al. (2009) Structural insights into eRF3 and stop codon recognition by eRF1. *Genes Dev.,* 23, 1106-1118.

Choudhary, C., Kumar, C., Gnad, F., Nielsen, M. L., Rehman, M., Walther, T. C., Olsen, J. V. & Mann, M. (2009). Lysine acetylation targets protein complexes and co-regulates major cellular functions. *Science,* 325, 834-840.

Demeshkina, N., Repkova, M., Ven'yaminova, A., Graifer, D. & Karpova, G. (2000) Nucleotides of 18S rRNA surrounding mRNA codons at the human ribosomal A, P and E sites, respectively: a cross-linking study with mRNA analogues carrying aryl azide group at either the uracil or the guanine residue. *RNA,* 6, 1727-1736.

Demeshkina, N., Laletina, E., Meschaninova, M., Ven'yaminova, A., Graifer, D. & Karpova, G. (2003a) Positioning of mRNA codons with respect to 18S rRNA at the P and E sites of human ribosome. *Biochim. Biophys. Acta,* 1627, 39-46.

Demeshkina, N. A., Laletina, E. S., Meschaninova, M. I., Repkova, M. N., Ven'yaminova, A. G., Graifer, D. M. & Karpova, G. G. (2003b) The mRNA codon environment at the P and E sites of human ribosomes deduced from photocrosslinking with pUUUGUU derivatives. *Mol. Biol. (Moscow),* 37, 132-139.

Demeshkina, N. A., Styazhkina, V. A., Bulygin, K. N., Repkova, M. N., Ven'yaminova, A. G. & Karpova, G. G. (2005) Template location on the human ribosome: environment of the mRNA nucleotide adjacent to the A-site codon on the 3'-side. *Russian J. Bioorgan. Chem.*, 31, 263–270.

Fukushi, S., Okada, M., Stahl, J., Kageyama, T., Hoshino, F. B. & Katayama, K. (2001) Ribosomal protein S5 interacts with the internal ribosomal entry site of hepatitis C virus. *J. Biol. Chem.*, 276, 20824-20826.

Garcia-Hernandez, M., Davies, E., Baskin, T. I. & Staswick, P. E. (1996) Association of plant p40 protein with ribosomes is enhanced when polyribosomes form during periods of active tissue growth. *Plant Physiol.,* 111, 559-568.

Ge, W., Wolf, A., Feng, T. et al. (2012) Oxygenase catalysed ribosome hydroxylation occurs in prokaryotes and humans. *Nat. Chem. Biol.,* 8, 960-962..

Graifer, D. M., Nekhai, S. Yu., Mundus, D. A., Fedorova, O. S. & Karpova, G. G. (1992) Interaction of human and Escherichia coli $tRNA^{Phe}$ with human 80S ribosomes in the presence of oligo- and polyuridylate template. *Biochim. Biophys. Acta*, 1171, 56-64.

Graifer, D., Molotkov, M., Styazhkina, V., Demeshkina, N., Bulygin, K., Eremina, A., Ivanov, A., Laletina, E., Ven'yaminova, A. & Karpova, G. (2004) Variable and conserved elements of human ribosomes surrounding the mRNA at the decoding and upstream sites. *Nucleic Acids Res.*, 32, 3282-3293.

Graifer, D., Zhigailov, A., Ven'yaminova, A., Malygin, A., Iskakov, B. & Karpova, G. (2012) 2'-OH of mRNA are critical for the binding of its codons at the 40S ribosomal P site but not at the mRNA entry site. *FEBS Lett.*, 586, 3731-3736.

Hellen, C. U. T. & Sarnow, P. (2001) Internal ribosome entry sites in eukaryotic mRNA molecules. *Genes Dev.,* 15, 1593-1612.

Hellen, C. U. (2009) IRES-induced conformational changes in the ribosome and the mechanism of translation initiation by internal ribosomal entry. *Biochim. Biophys. Acta*, 1789, 558-570.

Hountondji, C., Bulygin, K., Woisard, A., Tuffery, P., Créchet, J.-B., Pech, M., Nierhaus, K. H., Karpova, G. & Baouz, S. (2012) Lys53 of ribosomal protein L36AL and the CCA-end of a tRNA at the P/E hybrid site are in close proximity on the human ribosome. *ChemBiochem.,* 13, 1791–1797.

Khairulina, Yu. S., Molotkov, M. V., Bulygin, K. N., Graifer, D. M., Ven'yaminova, A. G., Frolova, L.Yu., Stahl, J. & Karpova, G. G. (2008) Protein S3 fragments neighboring mRNA during elongation and termination of translation on the human ribosome. *Russian J. Bioorgan. Chem.*, 34, 691–697.

Khairulina, J., Graifer, D., Bulygin, K., Ven'yaminova, A., Frolova, L. & Karpova, G. (2010) Eukaryote-specific motif of ribosomal protein S15 neighbors A site codon during elongation and termination of translation. *Biochimie,* 92, 820-825.

Kisselev, L. L., Ehrenberg, M. & Frolova, L. Yu. (2003) Termination of translation: interplay of mRNA, rRNAs and release factors. *EMBO J.*, 22, 175-182.

Klinge, S., Voigts-Hoffmann, F., Leibundgut, M., Arpagaus, S. & Ban, N. (2011) Crystal structure of the eukaryotic 60S ribosomal subunit in complex with initiation factor 6. *Science,* 334, 941-948.

Korostelev, A., Trakhanov, S., Laurberg, M. & Noller, H. F. (2006) Crystal structure of a 70S ribosome-tRNA complex reveals functional interactions and rearrangements. *Cell,* 126, 1065-1077.

Korostelev, A. A. (2011) Structural aspects of translation termination on the ribosome. *RNA,* 17, 1409–1421.

Kossinova, O. A., Malygin, A. A., Babaylova, E. S. & Karpova, G. G. (2008) Binding of human ribosomal protein p40 and its truncated mutants to the small ribosomal subunit. *Mol. Biol.* (Moscow), 42, 911–916.

Laletina, E., Graifer, D., Malygin, A., Ivanov, A., Shatsky, I. & Karpova, G. (2006) Proteins surrounding hairpin IIIe of the hepatitis C virus internal ribosome entry site on the human 40S ribosomal subunit. *Nucleic Acids Res.*, 34, 2027-2036.

Landry, D. M., Hertz, M. I. & Thompson, S. R. (2009) RPS25 is essential for translation initiation by the Dicistroviridae and hepatitis C viral IRESs. *Genes Dev.*, 23, 2753-2764.

Liang, H., Wong, J. Y., Bao, Q., Cavalcanti, A. R. O. & Landweber, L. F. (2005) Decoding region: analysis of eukaryotic release factor (eRF1) stop codon binding residues. *J. Mol. Evol.*, 60, 337-344.

Lill, R., Robertson, J. M. & Wintermeyer, W. (1986) Affinities of tRNA binding sites of ribosomes from Escherichia coli. *Biochemistry*, 25, 3245-3255.

Locker, N., Easton, L. E. & Lukavsky, P. J. (2007) HCV and CSFV IRES domain II mediate eIF2 release during 80S ribosome assembly. *EMBO J.*, 26, 795-805.

Louie, D. F., Resing, K. A., Lewis, T. S. & Ahn, N. G. (1996) Mass spectrometric analysis of 40 S ribosomal proteins from Rat-1 fibroblasts. *J. Biol. Chem.*, 271, 28189–28198.

Lukavsky, P. J. (2009) Structure and function of HCV IRES domains. *Virus Res.*, 139, 166-171.

Malygin, A. A., Bochkaeva, Z. V., Bondarenko, E. I., Kossinova, O. A., Loktev, V. B., Shatsky, I. N. & Karpova, G. G. (2009) Binding of the IRES of hepatitis C virus RNA to the 40S ribosomal subunit: role of p40. *Mol. Biol. (Moscow)*, 43, 997–1003.

Malygin, A. A. & Karpova, G. G. (2010) Site-specific cleavage of the 40S ribosomal subunit reveals eukaryote-specific ribosomal protein S28 in the subunit head. *FEBS Lett.*, 584, 4396-4400.

Malygin, A. A., Babaylova, E. S., Loktev, V. B. & Karpova, G. G. (2011) A region in the C-terminal domain of ribosomal protein SA required for binding of SA to the human 40S ribosomal subunit. *Biochimie,* 93, 612-617.

Malygin, A. A., Shatsky, I. N. & Karpova, G. G. (2012) Proteins of the human 40S ribosomal subunit involved in the hepatitis C IRES binding as revealed from fluorescent labelling. *Biochemistry (Moscow)*, 78, in press.

Moazed, D. & Noller, H. F. (1989) Interaction of tRNA with 23S rRNA in the ribosomal A, P, and E sites. *Cell*, 57, 585-597.

Molotkov, M., Graifer, D., Demeshkina, N., Repkova, M., Ven'yaminova, A. & Karpova, G. (2005) Arrangement of mRNA 3' of the A site codon on the human 80S ribosome. *RNA Biology*, 2, 63-69.

Molotkov, M. V., Graifer, D. M., Popugaeva, E. A., Bulygin, K. N., Meschaninova, M. I., Ven'yaminova, A. G. & Karpova, G. G. (2006) mRNA 3' of the A site bound codon is located close to protein S3 on the human 80S ribosome. *RNA Biology*, 3, 122-129.

Noller, H. F. (2006) Biochemical characterization of the ribosomal decoding site. *Biochimie.* 88, 935-941.

Ogle, J. M., Brodersen, D. E., Clemons, W. M. Jr., Tarry, M. J., Carter, A. P. & Ramakrishnan, V. (2001) Recognition of cognate transfer RNA by the 30S ribosomal subunit. *Science,* 292, 897-902.

Otto, G. A., Lukavsky, P. J., Lancaster, A. M., Sarnow, P. & Puglisi, J. D. (2002) Ribosomal proteins mediate the hepatitis C virus IRES-HeLa 40S interaction. *RNA*, 8, 913-923.

Otto, G. A. & Puglisi, J. D. (2004) The pathway of HCV IRES-mediated translation initiation. *Cell,* 119, 369–380.

Pestova, T. V., Shatsky, I. N., Fletcher, S. P., Jackson, R. J. & Hellen, C. U. T. (1998) A prokaryotic-like mode of cytoplasmic eukaryotic ribosome binding to the initiation codon during internal translation initiation of hepatitis C and classical swine fever virus RNAs. *Genes Dev.*, 12, 67-83.

Pisarev, A. V., Kolupaeva, V. G., Pisareva, V. P., Merrick, W. C., Hellen, C. U. T. & Pestova, T. V. (2006) Specific functional interaction of nucleotides at key –3 and +4 positions flanking the initiation codon with components of the mammalian 48S translation initiation complex. *Genes Dev.*, 20, 624-636.

Pisarev, A. V., Kolupaeva, V. G., Yusupov, M. M., Hellen, C. U. T. & Pestova, T. V. (2008) Ribosomal position and contacts of mRNA in eukaryotic translation initiation complexes. *EMBO J.*, 27, 1609-1621.

Potapov, A. P., Triana-Alonso, F. J. & Nierhaus, K. H. (1995) Ribosomal decoding processes at codons in the A or P sites depend differently on 2'-OH groups. *J. Biol. Chem.*, 270, 17680-17684.

Rabl, J., Leibundgut, M., Ataide, S.F., Haag, A. & Ban, N. (2011) Crystal structure of the eukaryotic 40S ribosomal subunit in complex with initiation factor 1. *Science,* 331, 730–736.

Ramakrishnan, V. (2002) Ribosome structure and the mechanism of translation. *Cell*, 108, 557-72.

Rheinberger, H.-J. (2004) A history of protein biosynthesis and ribosome research. In *Protein Synthesis and Ribosome Structure* (Nierhaus, K. H. & Wilson, D. N., eds), pp.1–51, Wiley-VCH Verlag.

Rodnina, M. & Wintermeyer, W. (2009) Recent mechanistic insights into eukaryotic ribosomes. *Curr. Opin. Cell. Biol.*, 21, 435–443.

Saenger, W. (1984) Principles of nucleic acid structure. In *Springer Advanced Texts in Chemistry* (Cantor, C. R., ed.), Chapter 18, New York, Berlin, Heidelberg, Tokyo, Springer-Verlag.

Schluenzen, F. Tocilj, A., Zarivach, R., Harms, J., Gluehmann, M., Janell, D., Bashan, A., Bartels, H., Agmon, I., Franceschi, F. & Yonath, A. (2000) Structure of functionally activated small ribosomal subunit at 3.3A° resolution. *Cell,* 102, 615–623.

Schmeing, T. M., Moore, P. B. & Steitz, T. A. (2003) Structures of deacylated tRNA mimics bound to the E site of the large ribosomal subunit. *RNA*, 9, 1345-1352.

Selmer, M., Dunham, C. M., Murphy IV F. M., Weixlbaumer, A., Petry, S., Kelley, A. C., Weir, J. R. & Ramakrishnan, V. (2006) Structure of the 70S ribosome complexed with mRNA and tRNA. *Science*, 313, 1935-1942.

Sharifulin, D., Khairulina, Yu., Ivanov, A., Meschaninova, M., Ven'yaminova, A., Graifer, D. & Karpova, G. (2012) A central fragment of ribosomal protein S26 containing the eukaryote-specific motif YxxPKxYxK is a key component of the ribosomal binding site of mRNA region 5' of the E site codon. *Nucleic Acids Res.,* 40, 3056-3065.

Shatsky, I. N., Dmitriev, S. E., Terenin, I. M. & Andreev, D. E. (2010) Cap- and IRES-independent scanning mechanism of translation initiation as an alternative to the concept of cellular IRESs. *Mol. Cells*, 30, 285-293.

Simonovic, M. & Steitz, T. A. (2008) Cross-crystal averaging reveals that the structure of the peptidyl-transferase center is the same in the 70S ribosome and the 50S subunit. *Proc. Natl. Acad. Sci. USA*, 105, 500-505.

Siridechadilok, B., Fraser, C. S., Hall, R. J., Doudna, J. A. & Nogales, E. (2005) Structural roles for human translation factor eIF3 in initiation of protein synthesis. *Science,* 310, 1513-1515.

Song, H., Mugnier, P., Webb, H. M., Evans, D. R., Tuite, M. F., Hemmings, B. A. & Barford, D. (2000) The crystal structure of human eukaryotic release factors eRF1 – mechanism of stop codon recognition and peptidyl-tRNA hydrolysis. *Cell,* 100, 311-321.

Spahn, C. M. T., Kieft, J. S., Grassucci, R. A., Penczek, P. A., Zhou, K., Doudna, J. A. & Frank, J. (2001) Hepatitis C virus IRES RNA-induced changes in the conformation of the 40S ribosomal subunit. *Science,* 291, 1959-1962.

Steitz, T. A. (2005) On the structural basis of peptide-bond formation and antibiotic resistance from atomic structures of the large ribosomal subunit. *FEBS Lett.,* 579, 955-958.

Steitz, T. A. (2008) A structural understanding of the dynamic ribosome machine. *Nat. Rev. Mol. Cell. Biol.*, 9, 242-253.

Taylor, D. J., Devkota, B., Huang, A. D., Topf, M., Narayanan, E., Sali, A., Harvey, S. C. & Frank, J. (2009) Comprehensive molecular structure of the eukaryotic ribosome. *Structure,* 17, 1591–1604.

Taylor, D., Unbehaun, A., Li, W., Das, S., Lei, J., Liao, H. Y., Grassucci, R. A., Pestova, T. V. & Frank, J. (2012) Cryo-EM structure of the mammalian eukaryotic release factor eRF1-eRF3-associated termination complex. *Proc. Natl. Acad. Sci. USA,*doi: 10.1073/pnas.1216730109.

Voigts-Hoffmann, F., Klinge, S. & Ban, N. (2012) Structural insights into eukaryotic ribosomes and the initiation of translation. *Curr. Opin. Struct. Biol.,* 22, 1-10.

Watanabe, S. (1972) Interaction of siomycin with the acceptor site of Escherichia coli ribosomes. *J. Mol. Biol.*, 67, 443-457.

Wimberly, B. T., Brodersen, D. E., Clemons, W. M. Jr., Morgan-Warren, R. J., Carter, A. P., Vonrhein, C., Hartsch, T. & Ramakrishnan, V. (2000) Structure of the 30S ribosomal subunit. *Nature,* 407, 327–339.

Wong, L. E., Li, Y., Pillay, S., Frolova, L. & Pervushin, K. (2012) Selectivity of stop codon recognition in translation termination is modulated by multiple conformations of GTS loop in eRF1. *Nucleic Acids Res.,* 40, 5751-5765.

Wower, J., Kirillov, S. V., Wower, I. K., Guven, S., Hixson, S. S. & Zimmermann, R. A. (2000) Transit of tRNA through the Escherichia coli ribosome. Cross-linking of the 3' end of tRNA to specific nucleotides of the 23 S ribosomal RNA at the A, P, and E sites. *J. Biol. Chem.*, 275, 37887-37894.

Yusupova, G. Zh., Yusupov, M. M., Cate, J. H. D. & Noller, H. F. (2001) The path of messenger RNA through the ribosome. *Cell,* 106, 233-241.

Yusupova, G., Jenner, L., Rees, B., Moras, D. & Yusupov, M. (2006) Structural basis for messenger RNA movement on the ribosome. *Nature,* 444, 391-394.

Zhao, W. D. & Wimmer, E. (2001) Genetic analysis of a poliovirus/hepatitis C virus chimera: new structure for domain II of the internal ribosomal entry site of hepatitis C virus. *J. Virol.*, 75, 3719-3730.

In: Ribosomes
Editors: Zhou Lin and Wang Liu
ISBN: 978-1-62417-698-2

Chapter 2

Ribosomal Alterations and Immune-Related Illness: Evidence of Links between Ribosomal Stress Signaling and Disease

Yuseok Moon *

Laboratory of Systems Mucosal Biomodulation, Department of Microbiology and Immunology, Pusan National University School of Medicine, Yangsan, South Korea
Research Institute for Basic Sciences, Pusan National University, Busan, South Korea

Abstract

Specific ribosome-directed stressors have the capacity to damage 28S ribosomal RNA and interfere with ribosomal functions during translation and cellular homeostasis. This can provoke so-called ribosomal stress responses that are closely associated with various disease processes including immunosuppressive or hypersensitivity-linked dysfunctions in humans and domestic animals. Since the primary toxic actions of most ribosomal stress-inducing agents generally inhibit global protein synthesis, highly dividing cells, including leukocytes and epithelial cells, are the most susceptible targets. Here, responses of immune cells and systems to acute and chronic exposure to ribosome-inactivating stress in various experimental models are discussed. This review will specifically focus on modulation of immune suppression and hypersensitivity diseases caused by ribosomal dysfunction. Evidence-based approaches characterizing ribosome-inactivating stressor and their implication as critical etiological factors of immune-associated diseases are also presented.

Keywords: Ribosome-inactivating stress, immune suppression, hypersensitivity

---

* To whom correspondence should be addressed: Yuseok Moon, Department of Microbiology and Immunology, Pusan National University School of medicine, Yangsan, 626-870 South Korea; Telephone: 82-51-510-8094, Fax: 82-55-382-8090, E-mail: moon@pnu.edu.

1. Introduction

Ribosomes are large complexes of RNA and protein which catalyze protein translation from individual amino using messenger RNA as a template. These complexes provide the platform where mRNA codons are deciphered with tRNA anticodons. Appropriate amino acids coupled with tRNA molecules are then delivered to the ribosome and incorporated into the growing polypeptide chain, thus forming a three-dimensional protein structure. Bacterial ribosomes have been extensively investigated as targets of many clinically used antibiotics that interfere with peptide bond formation (McCusker and Fujimori 2012, Wilson 2011). Moreover, specific ribosome-directed xenobiotics have the capacity to damage eukaryotic ribosomal RNA (rRNA) by inhibiting rRNA functioning during gene translation. Ribosome-inactivating proteins (RIPs) have been identified in various biological systems as natural products with rRNA N-glycosidase activity, particularly 28s rRNA cleavage, which leads to peptidyltransferase dysfunction and subsequent global translational arrest (Lacadena et al. 2007, Ng et al. 2010). Furthermore, other non-proteinaceous insults including ultraviolet (UV) irradiation, trichothecene mycotoxins, and anisomycin antibiotics also interfere with peptidyltransferase activity by directly or indirectly modifying 28s rRNA. These actions are collectively known as ribosome-inactivating stress or ribosomal stress although different cellular components are known to be major targets (Iordanov et al. 1998, Li and Pestka 2008). This type of toxic insult results in the expression of genes important for cellular homeostasis as well as a variety of pathogenic processes that modulate cell survival, proliferation, and differentiation (Laskin et al. 2002, Shifrin and Anderson 1999).

Body homeostatic integrity is widely defined as a balance between immune responses to external stresses and the counteracting immune tolerance. Depending on immunological imbalances, pathogenic pathways that predispose to diseases can manifest as an abnormally elevated immune response (hypersensitivity), decreased host resistance, and immune suppression. In mechanistic studies of ribosome-inactivating stress, both immune suppression and hypersensitivity have been observed in experimental models and clinical cases depending on insult exposure factors such as duration, doses, and frequency (Berek et al. 2002, Li et al. 2006a, Li et al. 2006b, Pestka and Smolinski 2005, Pestka et al. 2004, Zhou et al. 2005). Additionally, several epidemiological studies have uncovered links between ribosomal stress and human mucosal epithelial disorders ranging from acute mucosal inflammatory disease to epithelial malignancy (Luo et al. 1990, Abreu et al. 2002, Bhat et al. 1989). Ribosomal stresses trigger mucosal and systemic inflammation, and evoke the production of inflammatory mediators in epithelial and other immune-related cells (Zhou et al. 1999, Moon et al. 2008, An et al. 2005). In response to ribosomal insults, the most susceptible cells are also highly dividing leukocytes and epithelial cells.

The present review focuses on pathological and molecular events mediated by ribosomal insult that are associated with systemic and mucosal immune-related illnesses. In particular, molecular evidence demonstrating the association of ribosomal dysfunction with sepsis and mucosal hypersensitivity in the airways and intestinal tracts will be presented. These evidence-based approaches showing the impact of ribosome-inactivating stress indicate that ribosomal alterations are critical etiological factors of immune-associated diseases.

2. Ribosomal Insult and Systemic Inflammatory Responses

2.1. Ribosomal Insult and Pro-Inflammatory Signals

Sepsis is a systemic medical disorder which triggers a whole-body inflammatory responses known as systemic inflammatory response syndrome (SIRS) and is induced by an infection (Bossink et al. 1998, Opal 1998). Sepsis results in excessive immune responses against infection and leads to organ failure via ischemic injuries (Gosemann et al.). However, patients with sepsis are ultimately immunosuppressed as demonstrated by their development of immune tolerance and loss of ability to destroy pathogens (Meakins et al. 1977, Lederer et al. 1999, Puyana et al. 1998, Xu et al. 1998, Oberholzer et al. 2001, Docke et al. 1997).

Sepsis can be triggered by not only infection but also poisoning by pathogen-associated compounds including microbial toxins. Animals and humans with endotoxemia caused by bacterial endotoxins are in critical danger of developing sepsis even after microbial destruction by antibiotics (Giroir et al. 1992). Moreover, toxicants that induce ribosomal damage such as Shiga toxins, trichochecene mycotoxins, and RIPs can cause a multi-systemic sepsis-like syndrome (Islam and Pestka 2006, Lundeen et al. 1986, Valles et al. 2005, Wong et al. 2007a). Lethal high doses of ribosome-inactivating xenobiotics such as ricin lead to the accumulation of pro-inflammatory RNA transcripts and substantial damage to 28S rRNA in multiple organs including the lung, kidney, spleen, liver, and blood (Wong et al. 2007a).

Agents that induce ribosomal stress also trigger strong inflammatory responses in leukocytes by promoting gene expression of pro-inflammatory cytokines. This includes interleukin (IL)-8, tumor necrosis factor (TNF) α, IL-1β and IL-6 in macrophages, and IL-2, IL-4, and IL-5 in T cells (Chung et al. 2003, Jia et al. 2004, Li et al. 2000, Li et al. 1997, Ouyang et al. 1996, Wong et al. 1998). Mechanistically, pro-inflammatory gene expression is mostly mediated by mitogen-activated protein kinase (MAPK) in cells with altered ribosomal functions (Gray et al. 2008, Islam et al. 2006, Leyva-Illades et al. 2012, Moon et al. 2007, Wong et al. 2007b). The activated MAPK signaling cascade then induces transcriptional activation via different downstream transcription factors such as nuclear factor kappa B (NF-κB), activating factor 1 (AP-1), and early growth response gene 1 product (EGR-1) (Moon et al. 2007, Sakiri et al. 1998, Van De Walle et al. 2008, Wong et al. 2007b). Although many different stress-related mediators have been reported to be involved in ribosomal stress responses, MAPK signaling and subsequent activation of transcription factors are considered to be the main stress-signaling clusters that facilitate the gene expression of pro-inflammatory mediators (Iordanov et al. 1997, Iordanov et al. 1998).

2.2. Systemic Leukopenia Caused by Ribosomal Dysfunction in Host with Sepsis-Like Symptoms

In humans and animals suffering from sepsis, a number of central and peripheral lymphocytes are decreased. The degree of lymphocyte apoptosis is related to sepsis severity since monocytes and granulocytes from patients with sepsis predominantly display an anti-apoptotic gene expression profile whereas CD4+ T cells show a pro-apoptotic gene expression profile (Hattori et al. 2010, Hoogerwerf et al. 2010, Vaki et al. 2011, Wheeler 2009). Ribosomal insults activate c-Jun N-terminal kinase (JNK)/p38 MAPKs, well-known cellular components of pro-apoptotic signaling scaffolds. This induces rapid apoptosis, particularly in human T lymphocytes (Shifrin and Anderson 1999). Deoxynivalenol, a ribosomal stress agent, also promotes apoptosis in specific T and B cell subsets within the thymus, spleen, and Peyer's patches (Islam et al. 2003, Pestka et al. 1994, Uzarski et al. 2003). In addition, ribosomal insults act synergistically with bacterial endotoxin, leading to lymphoid atrophy and apoptosis in murine lymphoid organs where endotoxin exposure markedly amplifies the lymphoid toxicity and production of sepsis-related pro-inflammatory cytokines in a glucocorticoid-dependent manner (Islam et al. 2003, Islam et al. 2002, Islam and Pestka 2003, Zhou et al. 1999).

The immunosuppressive effects of ribosomal stress agents have been studied extensively in various animal models. These effects include leucopenia, decreased host resistance to pathogens, impaired delayed hypersensitivity, and altered antibody responses (Damian et al. 2011, Holladay et al. 1995, Minervini et al. 2005, Sauter et al. 2008, Svirshchevskaya et al. 2000, Venturini et al. 1996). As mentioned above, the most susceptible targets are lymphatic organs and epithelial tissues, including the thymus, bone marrow, spleen, mucosal lymph node, and intestinal mucosa, that contain rapidly dividing cells. The acute effects of lethal ribosomal stress on the immune system are similar to those seen with radiation exposure; thus, ribosomal stress is considered to have radiomimetic properties (Friend et al. 1983, Lorenzana et al. 1985, Lutsky and Mor 1981). In most experimental models, ribosomal stress-induced immunosuppression manifests as depressed activity of T lymphocytes and natural killer cells, which is linked to attenuated protection against tumor development and infection. Reduced activities of immune-related cells may be mechanistically due to direct cytotoxicity rather than the decreased production of immune mediators. However, the lethal effects of ribosomal stress on lymphocytes may also be associated with the up-regulated production of pro-inflammatory cytokines including TNFα (Korcheva et al. 2005, Korcheva et al. 2007, Lentz et al. 2010).

2.3. Crosstalk between Ribosomal Stress and Infectious Agents during Systemic Inflammatory Responses

Crosstalk between ribosomal stress and microbial infection has been relatively well-documented in different models (Ali-Vehmas et al. 1998, Boonchuvit et al. 1975, Ralph and Nakoinz 1973, Vandenbroucke et al. 2009). Susceptibility to microbial infection can be altered in hosts exposed to ribosomal stress. For example, ribosomal stress promotes the macrophage uptake of *Salmonella typhimurium* (Vandenbroucke et al. 2009). Enhanced uptake of bacteria coincides with marked macrophage F-actin re-organization due to the activation of extracellular signal-regulated protein kinase 1/2 (ERK1/2), indicating that ribosomal stress enhances susceptibility to microbial infection via the MAPK pathway. One recently published study also accounts for the link between ERK1/2 signaling and actin dynamics during phagocytosis (Mendoza et al. 2011). ERK1/2 directly phosphorylates the Wiskott–Aldrich syndrome protein (WASP)-verprolin-homologous protein 2 (WAVE2) regulatory complex, which activates actin polymerization. This type of coordinated signaling cascade linked to MAPKs is a general pattern of mammalian response to ribosomal stress (Berin et al. 2002, Wang et al. 2004).

Although the molecular mechanism underlying MAPK activation by ribosomal dysfunction has not been clearly addressed, one potent upstream sentinel, protein kinase R (PKR), may a play pivotal role in this process. PKR is a serine/threonine kinase which binds to and is activated by damaged ribosomes via interaction with two dsRNA-binding domains (Gray et al. 2008, Nallagatla et al. 2011). In addition to global translational arrest induced by PKR-phosphorylated eukaryotic initiation factor 2, PKR can form a signaling scaffold with damaged ribosomes for the direct binding and activation of MAPKs (Alisi et al. 2008, Iordanov et al. 2000, Zhang et al. 2009). In particular, bacterial lipopolysaccharide (LPS) is a chemical compound that plays a principal role in mediating immune and inflammatory responses to Gram-negative bacteria. Bacterial LPS can influence the magnitude of responses to xenobiotic agents (Brown et al. 1996, Kinser et al. 2002, Shaw et al. 2009) and has also been shown to synergistically enhance the cytotoxicity of ribosomal stress agents (Boonchuvit et al. 1975, Tai and Pestka 1988, Zhou et al. 2000). In response to co-exposure to LPS and ribosomal stress agents, most prominent molecular pattern is the super-induction of pro-inflammatory mediator expression, which is strongly associated with apoptosis and systemic sepsis-like syndrome due to ribosomal stress (Efrat et al. 1984, Jandhyala et al. 2010, Li et al. 1997, Wong et al. 2001).

3. Ribosomal Insult and Mucosal Inflammatory Responses

3.1. Effects of Ribosomal Stress on Mucosal Inflammation in Airways

Exposure to ribosome-inactivating stress in the airway is closely associated with severe epithelial inflammatory responses to acute lung injury and adult respiratory distress syndrome (Underwood et al. 1995a, Underwood et al. 1995b, Wong et al. 2007b). Pulmonary symptoms after ribosomal stress exposure include pulmonary edema, acute alveolitis, apoptosis, and massive infiltration of inflammatory cells (Wilhelmsen and Pitt 1996, Brown and White 1997, Tamura et al. 2002, Wong et al. 2007a). *Stachybotrys chartarum* is a well-known toxic black mold that has been evaluated as a strong etiological factor of pulmonary hemorrhage, known as sick-building syndrome (SBS), in infants (Pieckova and Jesenska 1999, Elidemir et al. 1999, Etzel et al. 1998, Montana et al. 1997, Novotny and Dixit 2000, Tripi et al. 2000, Cooley et al. 1998, Mahmoudi and Gershwin 2000). Exposure to this mold in homes enhances the development of allergic responses, resulting in respiratory dysfunction. Animals exposed to *Stachybotrys chartarum*-derived extracts have elevated levels of bronchoalveolar lavage fluid (BALF), eosinophils, lymphocytes, and pro-inflammatory cytokines such as IL-5 and IL-13 secreted by Th2 cells in BALF. These extracts also induce allergic asthma-like responses and hemorrhage in lungs of infants (Viana et al. 2002, Dearborn et al. 1999).

Asthma is a chronic respiratory disease characterized by pulmonary inflammatory responses such as mucus over-production and airway hyper-responsiveness to external antigens, leading to chest tightness, wheezing, and shortness of breath. Ribosome-inactivating macrocyclic trichothecenes such as satratoxin G (SG) produced by *Stachybotrys chartarum* trigger asthma-related respiratory immune responses (Amuzie et al. 2010, Gregory et al. 2004, Jarvis et al. 1998, Lichtenstein et al. 2010). Granulomatous inflammatory cells infiltrate into all compartments of the lung such as peribronchial, perivascular, and alveolar compartments during asthma-related respiratory illness. Macrophages are the main type of cells recovered by bronchoalveolar lavage (BAL) from both non-asthmatic and asthmatic individuals. Granulomas are predominantly macrophages (Eschenbacher and Gravelyn 1987). Chronically activated macrophages produce pro-inflammatory cytokines including IL-1, TNF-α, IL-6, and Granulocyte macrophage colony-stimulating factor (GM-CSF) which recruit inflammatory cells and prolong eosinophil survival (Gosset et al. 1992). Similar to other ribosome-inactivating trichothecens, SG activates the MAPK signaling pathway and induces leukocyte apoptosis, thereby altering airway defenses (Nusuetrong et al. 2005, Yang et al. 2000). Moreover, exposure of primary human airway epithelial cells to ribosomal stress activates the NF-κB pathway and stress-activated protein kinases (SAPKs), which promote the induction of pro-inflammatory gene expression. The resulting precursor cytokines are then processed via inflammasome-mediated cleavage (Jandhyala et al. 2012, Vyleta et al. 2012, Wong et al. 2007b). Ribosome-inactivating stress also increases total IgE concentrations about 10-fold and augments the levels of eosinophils, neutrophils, and lymphocytes in lavage fluid (Underwood et al. 1995b). Studies in rodents and nonhuman primates have demonstrated that exposure to ribosomal stress via aerosol delivery elevates the production of central pro-inflammatory cytokines such as IL-1, which mediate acute lung injury and the development of symptoms resembling acute respiratory distress syndrome (Amuzie et al. 2008, Korcheva et al. 2007, Lichtenstein et al. 2010, Lindauer et al. 2009). In addition to cytokine production, other pro-inflammatory mediators including prostaglandins and reactive oxygen species contribute to epithelial inflammatory responses, tissue injuries, and even fetal mortality associated with damaged placenta (Bitzan et al. 1998, Burdet et al. 2010, Williams et al. 2000). Considering the ability of ribosomal stress to enhance the production and activity of different pro-inflammatory mediators, commonly shared pro-inflammatory signaling pathways and networks need to be further investigated.

3.2. Effects of Ribosomal Stress on Mucosal Inflammatory Signals in the Gastrointestinal Tract

Human inflammatory bowel disease (IBD) is a gastrointestinal disorder characterized by chronic hypersensitive immune activation and accompanying ulcerative lesions. IBD can be classified into two representative types: ulcerative colitis (UC) and Crohn's disease (CD). CD features deep transmural ulcerative inflammation with granulomas. This condition can affect any part of the gastrointestinal tract although it most commonly occurs the the distal ileum and cecum. Patients with UC have more superficial ulceration limited to the colon, particularly the distal colon and rectum.

Chemically-induced ribosomal dysfunction alters intestinal mucosal integrity by interfering with transepithelial resistance, epithelial differentiation, and nutrient absorption; these disruptions are associated with anorexia and weight loss (Bouhet and Oswald 2005, Calvert et al. 2005, Maresca et al. 2002, Sergent et al. 2006). The epithelia is the first sentinel against mucosal ribosomal stress, and transmits the danger signals to the underlying lymphocytes and antigen-presenting cells by releasing pro-inflammatory chemokines such as

IL-8 (Gobert et al. 2007, Jeong et al. 2010, Park et al. 2010, Van De Walle et al. 2010). As previously suggested, the most characteristic cellular responses to ribosomal stress are the activation of MAPK (Jandhyala et al. 2008, Pestka et al. 2005, Shifrin and Anderson 1999) and subsequent activation of downstream transcription factors such as NF-κB, AP-1, CCAAT enhancer binding protein (C/EBP), cyclic AMP response element binding protein (CREB), and early growth response 1 (EGR1), all of which are critical for the transcriptional activation of pro-inflammatory mediators (Jandhyala et al. 2010, Moon et al. 2008, Sakiri et al. 1998, Shifrin and Anderson 1999, Van De Walle et al. 2010). Most of all, ribosomal stress agents can activate the central pro-inflammatory transcription factor NF-κB in some cell types such as monocyte-derived cells while extended exposure to the toxins suppresses this signal (Hauf and Chakraborty 2003, Sugimoto et al. 2000).

However, induction of epithelial pro-inflammatory cytokines by these compounds, including trichothecenes or anisomycin, is independent of NF-κB signaling in the intestinal epithelia (Moon et al. 2008, Park et al. 2010). Instead of NF-κB, EGR1 is involved in the expression of pro-inflammatoryIL-8 in intestinal epithelial cells exposed to ribosomal stress. Although there is no EGR1 binding motif in the IL-8 gene promoter, EGR1 can function as an inducer of cytokine expression that acts cooperatively with AP-1 (Levkovitz and Baraban 2001, Levkovitz and Baraban 2002). EGR1-mediated IL-8 production could be associated with three central transcription factors, NF-κB, C/EBP, and AP-1, since these components also contribute to ribosomal stress-induced cytokine transcription activation. *EGR1* expression can be induced by a range of stimuli, which in many cases overlap with those known to be capable of activating NF-κB signals (Ahmed et al. 1997, Milbrandt 1987, Yan et al. 1998). NF-κB and *EGR1* have been shown to cooperatively stimulate the transcription of pro-inflammatory cytokine genes (Cogswell et al. 1997, Ma et al. 2009, Shin et al. 2010) although a report showed that *EGR1* specifically represses NF-κB transcriptional activity in a feedback loop by preventing NF-κB interaction with its promoter target elements (Chapman and Perkins 2000). In addition, it has been found that AP-1 is a positive downstream of EGR1 (Levkovitz and Baraban 2002). Therefore, the cooperation between EGR-1 and other pro-inflammatory transcription factors may be closely associated with the promotion of cytokine gene transcription despite partial contributions of individual cis-acting elements.

In addition to MAPK-driven transcriptional regulation, post-transcriptional modification such as mRNA stabilization may enhance cytokine production in response to ribosomal insults (Choi et al. 2009, Moon et al. 2007, Park et al. 2010). Mechanistically, HuR/Elav-like RNA binding protein 1 positively regulates the stability of mRNA transcripts containing AU-rich elements (AREs) including ones encoding pro-inflammatory cytokines (Fan et al. 2011, Feng et al. 2011, Park et al. 2012). HuR protein then binds to the AREs in the 3' untranslated region (3'UTR) of target mRNA molecules in the nucleus and translocates into the cytoplasm for translation. Cytosolic translocation of the HuR protein is also initiated by mucosal ribosomal stress and ultimately stabilizes cytokine mRNA (Choi et al. 2009, Park et al. 2010).

In summary, transcriptional activation of cytokine genes is promoted by either MAPK-activated NF-κB or other transcription factors independent of NF-κB. Some types of ribosomal stress can elevate inflammasome signals which play important roles in processing precursor cytokines. Moreover, post-transcriptional modification such as mRNA stabilization also increases cytokine production in response to ribosomal insults.

Conclusion

Accumulating evidence has indicated that ribosomal stress is an etiological factor of systemic and mucosal immune-related diseases. Ribosomal insults cause a multi-systemic sepsis-like syndrome, which manifests as leucopenia including lymphoid atrophy and apoptosis due directly to cytotoxicity rather than decreased production of immune mediators. However, the lethal effects of ribosomal stress in lymphoid tissues can be associated with up-regulated production of some pro-inflammatory cytokines. Susceptibility to microbial infection is altered in hosts exposed to ribosomal stress with marked macrophage F-actin re-organization due to MAPK activation. Moreover, co-exposure to bacterial endotoxins and ribosomal stress triggers the super-induction of pro-inflammatory mediator expression that is strongly associated with systemic sepsis-like syndrome. Exposure to airway ribosome-inactivating stress is closely linked to severe airway inflammatory responses with acute lung injury and respiratory distress syndrome. In particular, ribosome-inactivating macrocyclic trichothecenes produced by *Stachybotrys chartarum* induce asthma-related respiratory immune response. Ribosomal stress alters intestinal mucosal integrity by interfering with transepithelial resistance, epithelial differentiation, and nutrient absorption, which may lead to anorexia and weight loss. In response to mucosal ribosomal stress, the epithelia act as the first defensive sentinel and transmit signals to the underlying lymphocytes and antigen-presenting cells. The transcription of cytokine genes can be activated by either MAPK-linked NF-κB or other transcription factors. Some types of ribosomal stress can induce inflammasome signals which play central roles in processing precursor cytokines. Moreover, post-transcriptional modifications such as mRNA stabilization may contribute to increased cytokine production following ribosomal insult. Despite these finding from the literature, further studies of the mediators discussed in the present review need to be performed to better assess their activities in humans, and compare their effects under normal and pathologic conditions. More systematic approaches involving translational medicine should be established for improved treatment and prevention of environmentally linked immune disorders in humans.

Acknowledgment

This work was supported by the Basic Science Research Program through the National Research Foundation of Korea funded by the Ministry of Education, Science, and Technology (grant 2012R1A1A2005837).

References

Abreu, M.T., Arnold, E.T., Thomas, L.S., Gonsky, R., Zhou, Y., Hu, B. & Arditi, M. (2002). Tlr4 And Md-2 Expression Is Regulated By Immune-Mediated Signals In Human Intestinal Epithelial Cells. *J Biol Chem,* Jun 7. 277: 20431-20437.

Ahmed, M.M., Sells, S.F., Venkatasubbarao, K., Fruitwala, S.M., Muthukkumar, S., Harp, C., Mohiuddin, M. & Rangnekar, V.M. (1997). Ionizing Radiation-Inducible Apoptosis In The Absence Of P53 Linked To Transcription Factor Egr-1. *J Biol Chem,* Dec 26. 272: 33056-33061.

Ali-Vehmas, T., Rizzo, A., Westermarck, T. & Atroshi, F. (1998). Measurement Of Antibacterial Activities Of T-2 Toxin, Deoxynivalenol, Ochratoxin A, Aflatoxin B1 And Fumonisin B1 Using Microtitration Tray-Based Turbidimetric Techniques. *Zentralbl Veterinarmed A,* Oct. 45: 453-458.

Alisi, A., Spaziani, A., Anticoli, S., Ghidinelli, M. & Balsano, C. (2008). Pkr Is A Novel Functional Direct Player That Coordinates Skeletal Muscle Differentiation Via P38mapk/Akt Pathways. *Cell Signal,* Mar. 20: 534-542.

Amuzie, C.J., Harkema, J.R. & Pestka, J.J. (2008). Tissue Distribution And Proinflammatory Cytokine Induction By The Trichothecene Deoxynivalenol In The Mouse: Comparison Of Nasal Vs. Oral Exposure. *Toxicology,* Jun 3. 248: 39-44.

Amuzie, C.J., Islam, Z., Kim, J.K., Seo, J.H. & Pestka, J.J. (2010). Kinetics Of Satratoxin G Tissue Distribution And Excretion Following Intranasal Exposure In The Mouse. *Toxicol Sci,* Aug. 116: 433-440.

An, H., Xu, H., Zhang, M., Zhou, J., Feng, T., Qian, C., Qi, R. & Cao, X. (2005). Src Homology 2 Domain-Containing Inositol-5-Phosphatase 1 (Ship1) Negatively Regulates Tlr4-Mediated Lps Response Primarily Through A Phosphatase Activity- And Pi-3k-Independent Mechanism. *Blood,* Jun 15. 105: 4685-4692.

Berek, L., Petri, I.B., Mesterhazy, A. & Teren, J. (2002). The Immunosuppressive Effect Of Fusarium Mycotoxin As A Function Of Hla Antigens. *Acta Microbiol Immunol Hung,* 49: 99-104.

Berin, M.C., Darfeuille-Michaud, A., Egan, L.J., Miyamoto, Y. & Kagnoff, M.F. (2002). Role Of Ehec O157:H7 Virulence Factors In The Activation Of Intestinal Epithelial Cell Nf-Kappab And Map Kinase Pathways And The Upregulated Expression Of Interleukin 8. *Cell Microbiol,* Oct. 4: 635-648.

Bhat, R.V., Beedu, S.R., Ramakrishna, Y. & Munshi, K.L. (1989). Outbreak Of Trichothecene Mycotoxicosis Associated With Consumption Of Mould-Damaged Wheat Production In Kashmir Valley, India. *Lancet,* Jan 7. 1: 35-37.

Bitzan, M.M., Wang, Y., Lin, J. & Marsden, P.A. (1998). Verotoxin And Ricin Have Novel Effects On Preproendothelin-1 Expression But Fail To Modify Nitric Oxide Synthase (Ecnos) Expression And No Production In Vascular Endothelium. *J Clin Invest,* Jan 15. 101: 372-382.

Boonchuvit, B., Hamilton, P.B. & Burmeister, H.R. (1975). Interaction Of T-2 Toxin With Salmonella Infections Of Chickens. *Poult Sci,* Sep. 54: 1693-1696.

Bossink, A.W., Groeneveld, J., Hack, C.E. & Thijs, L.G. (1998). Prediction Of Mortality In Febrile Medical Patients: How Useful Are Systemic Inflammatory Response Syndrome And Sepsis Criteria? *Chest,* Jun. 113: 1533-1541.

Bouhet, S. & Oswald, I.P. (2005). The Effects Of Mycotoxins, Fungal Food Contaminants, On The Intestinal Epithelial Cell-Derived Innate Immune Response. *Vet Immunol Immunopathol,* Oct 18. 108: 199-209.

Brown, A.P., Schultze, A.E., Holdan, W.L., Buchweitz, J.P., Roth, R.A. & Ganey, P.E. (1996). Lipopolysaccharide-Induced Hepatic Injury Is Enhanced By Polychlorinated Biphenyls. *Environ Health Perspect,* Jun. 104: 634-640.

Brown, R.F. & White, D.E. (1997). Ultrastructure Of Rat Lung Following Inhalation Of Ricin Aerosol. *Int J Exp Pathol,* Aug. 78: 267-276.

Burdet, J., Zotta, E., Cella, M., Franchi, A.M. & Ibarra, C. (2010). Role Of Nitric Oxide In Shiga Toxin-2-Induced Premature Delivery Of Dead Fetuses In Rats. *Plos One,* 5: E15127.

Calvert, T.W., Aidoo, K.E., Candlish, A.G. & Fuat, A.R. (2005). Comparison Of In Vitro Cytotoxicity Of Fusarium Mycotoxins,Deoxynivalenol, T-2 Toxin And Zearalenone On Selected Human Epithelial Cell Lines. *Mycopathologia,* Apr. 159: 413-419.

Chapman, N.R. & Perkins, N.D. (2000). Inhibition Of The Rela(P65) Nf-Kappab Subunit By Egr-1. *J Biol Chem,* Feb 18. 275: 4719-4725.

Choi, H.J., Yang, H., Park, S.H. & Moon, Y. (2009). Hur/Elavl1 Rna Binding Protein Modulates Interleukin-8 Induction By Muco-Active Ribotoxin Deoxynivalenol. *Toxicol Appl Pharmacol,* Oct 1. 240: 46-54.

Chung, Y.J., Zhou, H.R. & Pestka, J.J. (2003). Transcriptional And Posttranscriptional Roles For P38 Mitogen-Activated Protein Kinase In Upregulation Of Tnf-Alpha Expression By Deoxynivalenol (Vomitoxin). *Toxicol Appl Pharmacol,* Dec 1. 193: 188-201.

Cogswell, P.C., Mayo, M.W. & Baldwin, A.S., Jr. (1997). Involvement Of Egr-1/Rela Synergy In Distinguishing T Cell Activation From Tumor Necrosis Factor-Alpha-Induced Nf-Kappa B1 Transcription. *J Exp Med,* Feb 3. 185: 491-497.

Cooley, J.D., Wong, W.C., Jumper, C.A. & Straus, D.C. (1998). Correlation Between The Prevalence Of Certain Fungi And Sick Building Syndrome. *Occup Environ Med,* Sep. 55: 579-584.

Damian, D.L., Matthews, Y.J., Phan, T.A. & Halliday, G.M. (2011). An Action Spectrum For Ultraviolet Radiation-Induced Immunosuppression In Humans. *Br J Dermatol,* Mar. 164: 657-659.

Dearborn, D.G., Yike, I., Sorenson, W.G., Miller, M.J. & Etzel, R.A. (1999). Overview Of Investigations Into Pulmonary Hemorrhage Among Infants In Cleveland, Ohio. *Environ Health Perspect,* Jun. 107 Suppl 3: 495-499.

Docke, W.D., Randow, F., Syrbe, U., Krausch, D., Asadullah, K., Reinke, P., Volk, H.D. & Kox, W. (1997). Monocyte Deactivation In Septic Patients: Restoration By Ifn-Gamma Treatment. *Nat Med,* Jun. 3: 678-681.

Efrat, S., Zelig, S., Yagen, B. & Kaempfer, R. (1984). Superinduction Of Human Interleukin-2 Messenger Rna By Inhibitors Of Translation. *Biochem Biophys Res Commun,* Sep 17. 123: 842-848.

Elidemir, O., Colasurdo, G.N., Rossmann, S.N. & Fan, L.L. (1999). Isolation Of Stachybotrys From The Lung Of A Child With Pulmonary Hemosiderosis. *Pediatrics,* Oct. 104: 964-966.

Eschenbacher, W.L. & Gravelyn, T.R. (1987). A Technique For Isolated Airway Segment Lavage. *Chest,* Jul. 92: 105-109.

Etzel, R.A., Montana, E., Sorenson, W.G., Kullman, G.J., Allan, T.M., Dearborn, D.G., Olson, D.R., Jarvis, B.B. & Miller, J.D. (1998). Acute Pulmonary Hemorrhage In Infants Associated With Exposure To Stachybotrys Atra And Other Fungi. *Arch Pediatr Adolesc Med,* Aug. 152: 757-762.

Fan, J., Ishmael, F.T., Fang, X., Myers, A., Cheadle, C., Huang, S.K., Atasoy, U., Gorospe, M. & Stellato, C. (2011). Chemokine Transcripts As Targets Of The Rna-Binding Protein Hur In Human Airway Epithelium. *J Immunol,* Feb 15. 186: 2482-2494.

Feng, S., Chen, W., Cao, D., Bian, J., Gong, F.Y., Cheng, W., Cheng, S., Xu, Q., Hua, Z.C. & Yin, W. (2011). Involvement Of Na(+), K (+)-Atpase And Its Inhibitors In Hur-Mediated Cytokine Mrna Stabilization In Lung Epithelial Cells. *Cell Mol Life Sci,* Jan. 68: 109-124.

Friend, S.C., Hancock, D.S., Schiefer, H.B. & Babiuk, L.A. (1983). Experimental T-2 Toxicosis In Sheep. *Can J Comp Med,* Jul. 47: 291-297.

Giroir, B.P., Johnson, J.H., Brown, T., Allen, G.L. & Beutler, B. (1992). The Tissue Distribution Of Tumor Necrosis Factor Biosynthesis During Endotoxemia. *J Clin Invest,* Sep. 90: 693-698.

Gobert, A.P., Vareille, M., Glasser, A.L., Hindre, T., De Sablet, T. & Martin, C. (2007). Shiga Toxin Produced By Enterohemorrhagic Escherichia Coli Inhibits Pi3k/Nf-Kappab Signaling Pathway In Globotriaosylceramide-3-Negative Human Intestinal Epithelial Cells. *J Immunol,* Jun 15. 178: 8168-8174.

Gosemann, J.H., Van Griensven, M., Barkhausen, T., Kobbe, P., Thobe, B.M., Haasper, C., Pape, H.C., Krettek, C., Hildebrand, F. & Frink, M. Tlr4 Influences The Humoral And Cellular Immune Response During Polymicrobial Sepsis. *Injury,* Oct. 41: 1060-1067.

Gosset, P., Tsicopoulos, A., Wallaert, B., Joseph, M., Capron, A. & Tonnel, A.B. (1992). Tumor Necrosis Factor Alpha And Interleukin-6 Production By Human Mononuclear Phagocytes From Allergic Asthmatics After Ige-Dependent Stimulation. *Am Rev Respir Dis,* Sep. 146: 768-774.

Gray, J.S., Bae, H.K., Li, J.C., Lau, A.S. & Pestka, J.J. (2008). Double-Stranded Rna-Activated Protein Kinase Mediates Induction Of Interleukin-8 Expression By Deoxynivalenol, Shiga Toxin 1, And Ricin In Monocytes. *Toxicol Sci,* Oct. 105: 322-330.

Gregory, L., Pestka, J.J., Dearborn, D.G. & Rand, T.G. (2004). Localization Of Satratoxin-G In Stachybotrys Chartarum Spores And Spore-Impacted Mouse Lung Using Immunocytochemistry. *Toxicol Pathol,* Jan-Feb. 32: 26-34.

Hattori, Y., Takano, K., Teramae, H., Yamamoto, S., Yokoo, H. & Matsuda, N. (2010). Insights Into Sepsis Therapeutic Design Based On The Apoptotic Death Pathway. *J Pharmacol Sci,* 114: 354-365.

Hauf, N. & Chakraborty, T. (2003). Suppression Of Nf-Kappa B Activation And Proinflammatory Cytokine Expression By Shiga Toxin-Producing Escherichia Coli. *J Immunol,* Feb 15. 170: 2074-2082.

Holladay, S.D., Smith, B.J. & Luster, M.I. (1995). B Lymphocyte Precursor Cells Represent Sensitive Targets Of T2 Mycotoxin Exposure. *Toxicol Appl Pharmacol,* Apr. 131: 309-315.

Hoogerwerf, J.J., Van Zoelen, M.A., Wiersinga, W.J., Van 'T Veer, C., De Vos, A.F., De Boer, A., Schultz, M.J., Hooibrink, B., De Jonge, E. & Van Der Poll, T. (2010). Gene Expression Profiling Of Apoptosis Regulators In Patients With Sepsis. *J Innate Immun,* 2: 461-468.

Iordanov, M.S., Paranjape, J.M., Zhou, A., Wong, J., Williams, B.R., Meurs, E.F., Silverman, R.H. & Magun, B.E. (2000). Activation Of P38 Mitogen-Activated Protein Kinase And C-Jun Nh(2)-Terminal Kinase By Double-Stranded Rna And Encephalomyocarditis Virus: Involvement Of Rnase L, Protein Kinase R, And Alternative Pathways. *Mol Cell Biol,* Jan. 20: 617-627.

Iordanov, M.S., Pribnow, D., Magun, J.L., Dinh, T.H., Pearson, J.A., Chen, S.L. & Magun, B.E. (1997). Ribotoxic Stress Response: Activation Of The Stress-Activated Protein Kinase Jnk1 By Inhibitors Of The Peptidyl Transferase Reaction And By Sequence-Specific Rna Damage To The Alpha-Sarcin/Ricin Loop In The 28s Rrna. *Mol Cell Biol,* Jun. 17: 3373-3381.

Iordanov, M.S., Pribnow, D., Magun, J.L., Dinh, T.H., Pearson, J.A. & Magun, B.E. (1998). Ultraviolet Radiation Triggers The Ribotoxic Stress Response In Mammalian Cells. *J Biol Chem,* Jun 19. 273: 15794-15803.

Islam, Z., Gray, J.S. & Pestka, J.J. (2006). P38 Mitogen-Activated Protein Kinase Mediates Il-8 Induction By The Ribotoxin Deoxynivalenol In Human Monocytes. *Toxicol Appl Pharmacol,* Jun 15. 213: 235-244.

Islam, Z., King, L.E., Fraker, P.J. & Pestka, J.J. (2003). Differential Induction Of Glucocorticoid-Dependent Apoptosis In Murine Lymphoid Subpopulations In Vivo Following Coexposure To Lipopolysaccharide And Vomitoxin (Deoxynivalenol). *Toxicol Appl Pharmacol,* Mar 1. 187: 69-79.

Islam, Z., Moon, Y.S., Zhou, H.R., King, L.E., Fraker, P.J. & Pestka, J.J. (2002). Endotoxin Potentiation Of Trichothecene-Induced Lymphocyte Apoptosis Is Mediated By Up-Regulation Of Glucocorticoids. *Toxicol Appl Pharmacol,* Apr 1. 180: 43-55.

Islam, Z. & Pestka, J.J. (2003). Role Of Il-1(Beta) In Endotoxin Potentiation Of Deoxynivalenol-Induced Corticosterone Response And Leukocyte Apoptosis In Mice. *Toxicol Sci,* Jul. 74: 93-102.

Islam, Z. & Pestka, J.J. (2006). Lps Priming Potentiates And Prolongs Proinflammatory Cytokine Response To The Trichothecene Deoxynivalenol In The Mouse. *Toxicol Appl Pharmacol,* Feb 15. 211: 53-63.

Jandhyala, D.M., Ahluwalia, A., Obrig, T. & Thorpe, C.M. (2008). Zak: A Map3kinase That Transduces Shiga Toxin- And Ricin-Induced Proinflammatory Cytokine Expression. *Cell Microbiol,* Jul. 10: 1468-1477.

Jandhyala, D.M., Rogers, T.J., Kane, A., Paton, A.W., Paton, J.C. & Thorpe, C.M. (2010). Shiga Toxin 2 And Flagellin From Shiga-Toxigenic Escherichia Coli Superinduce Interleukin-8 Through Synergistic Effects On Host Stress-Activated Protein Kinase Activation. *Infect Immun,* Jul. 78: 2984-2994.

Jandhyala, D.M., Thorpe, C.M. & Magun, B. (2012). Ricin And Shiga Toxins: Effects On Host Cell Signal Transduction. *Curr Top Microbiol Immunol,* 357: 41-65.

Jarvis, B.B., Sorenson, W.G., Hintikka, E.L., Nikulin, M., Zhou, Y., Jiang, J., Wang, S., Hinkley, S., Etzel, R.A. & Dearborn, D. (1998). Study Of Toxin Production By Isolates Of Stachybotrys Chartarum And Memnoniella Echinata Isolated During A Study Of Pulmonary Hemosiderosis In Infants. *Appl Environ Microbiol,* Oct. 64: 3620-3625.

Jeong, K.I., Zhang, Q., Nunnari, J. & Tzipori, S. (2010). A Piglet Model Of Acute Gastroenteritis Induced By Shigella Dysenteriae Type 1. *J Infect Dis,* Mar 15. 201: 903-911.

Jia, Q., Zhou, H.R., Bennink, M. & Pestka, J.J. (2004). Docosahexaenoic Acid Attenuates Mycotoxin-Induced Immunoglobulin A Nephropathy, Interleukin-6 Transcription, And Mitogen-Activated Protein Kinase Phosphorylation In Mice. *J Nutr,* Dec. 134: 3343-3349.

Kinser, S., Copple, B.L., Roth, R.A. & Ganey, P.E. (2002). Enhancement Of Allyl Alcohol Hepatotoxicity By Endotoxin Requires Extrahepatic Factors. *Toxicol Sci,* Oct. 69: 470-481.

Korcheva, V., Wong, J., Corless, C., Iordanov, M. & Magun, B. (2005). Administration Of Ricin Induces A Severe Inflammatory Response Via Nonredundant Stimulation Of Erk, Jnk, And P38 Mapk And Provides A Mouse Model Of Hemolytic Uremic Syndrome. *Am J Pathol,* Jan. 166: 323-339.

Korcheva, V., Wong, J., Lindauer, M., Jacoby, D.B., Iordanov, M.S. & Magun, B. (2007). Role Of Apoptotic Signaling Pathways In Regulation Of Inflammatory Responses To Ricin In Primary Murine Macrophages. *Mol Immunol,* Apr. 44: 2761-2771.

Lacadena, J., Alvarez-Garcia, E., Carreras-Sangra, N., Herrero-Galan, E., Alegre-Cebollada, J., Garcia-Ortega, L., Onaderra, M., Gavilanes, J.G. & Martinez Del Pozo, A. (2007). Fungal Ribotoxins: Molecular Dissection Of A Family Of Natural Killers. *Fems Microbiol Rev,* Mar. 31: 212-237.

Laskin, J.D., Heck, D.E. & Laskin, D.L. (2002). The Ribotoxic Stress Response As A Potential Mechanism For Map Kinase Activation In Xenobiotic Toxicity. *Toxicol Sci,* Oct. 69: 289-291.

Lederer, J.A., Rodrick, M.L. & Mannick, J.A. (1999). The Effects Of Injury On The Adaptive Immune Response. *Shock,* Mar. 11: 153-159.

Lentz, E.K., Cherla, R.P., Jaspers, V., Weeks, B.R. & Tesh, V.L. (2010). Role Of Tumor Necrosis Factor Alpha In Disease Using A Mouse Model Of Shiga Toxin-Mediated Renal Damage. *Infect Immun,* Sep. 78: 3689-3699.

Levkovitz, Y. & Baraban, J.M. (2001). A Dominant Negative Inhibitor Of The Egr Family Of Transcription Regulatory Factors Suppresses Cerebellar Granule Cell Apoptosis By Blocking C-Jun Activation. *J Neurosci,* Aug 15. 21: 5893-5901.

Levkovitz, Y. & Baraban, J.M. (2002). A Dominant Negative Egr Inhibitor Blocks Nerve Growth Factor-Induced Neurite Outgrowth By Suppressing C-Jun Activation: Role Of An Egr/C-Jun Complex. *J Neurosci,* May 15. 22: 3845-3854.

Leyva-Illades, D., Cherla, R.P., Lee, M.S. & Tesh, V.L. (2012). Regulation Of Cytokine And Chemokine Expression By The Ribotoxic Stress Response Elicited By Shiga Toxin Type 1 In Human Macrophage-Like Thp-1 Cells. *Infect Immun,* Jun. 80: 2109-2120.

Li, M., Cuff, C.F. & Pestka, J.J. (2006a). T-2 Toxin Impairment Of Enteric Reovirus Clearance In The Mouse Associated With Suppressed Immunoglobulin And Ifn-Gamma Responses. *Toxicol Appl Pharmacol,* Aug 1. 214: 318-325.

Li, M., Harkema, J.R., Islam, Z., Cuff, C.F. & Pestka, J.J. (2006b). T-2 Toxin Impairs Murine Immune Response To Respiratory Reovirus And Exacerbates Viral Bronchiolitis. *Toxicol Appl Pharmacol,* Nov 15. 217: 76-85.

Li, M. & Pestka, J.J. (2008). Comparative Induction Of 28s Ribosomal Rna Cleavage By Ricin And The Trichothecenes Deoxynivalenol And T-2 Toxin In The Macrophage. *Toxicol Sci,* Sep. 105: 67-78.

Li, S., Ouyang, Y., Yang, G.H. & Pestka, J.J. (2000). Modulation Of Transcription Factor Ap-1 Activity In Murine El-4 Thymoma Cells By Vomitoxin (Deoxynivalenol). *Toxicol Appl Pharmacol,* Feb 15. 163: 17-25.

Li, S., Ouyang, Y.L., Dong, W. & Pestka, J.J. (1997). Superinduction Of Il-2 Gene Expression By Vomitoxin (Deoxynivalenol) Involves Increased Mrna Stability. *Toxicol Appl Pharmacol,* Dec. 147: 331-342.

Lichtenstein, J.H., Molina, R.M., Donaghey, T.C., Amuzie, C.J., Pestka, J.J., Coull, B.A. & Brain, J.D. (2010). Pulmonary Responses To Stachybotrys Chartarum And Its Toxins: Mouse Strain Affects Clearance And Macrophage Cytotoxicity. *Toxicol Sci,* Jul. 116: 113-121.

Lindauer, M.L., Wong, J., Iwakura, Y. & Magun, B.E. (2009). Pulmonary Inflammation Triggered By Ricin Toxin Requires Macrophages And Il-1 Signaling. *J Immunol,* Jul 15. 183: 1419-1426.

Lorenzana, R.M., Beasley, V.R., Buck, W.B. & Ghent, A.W. (1985). Experimental T-2 Toxicosis In Swine. Ii. Effect Of Intravascular T-2 Toxin On Serum Enzymes And Biochemistry, Blood Coagulation, And Hematology. *Fundam Appl Toxicol,* Oct. 5: 893-901.

Lundeen, G.R., Poppenga, R.H., Beasley, V.R., Buck, W.B., Tranquilli, W.J. & Lambert, R.J. (1986). Systemic Distribution Of Blood Flow During T-2 Toxin Induced Shock In Swine. *Fundam Appl Toxicol,* Aug. 7: 309-323.

Luo, Y., Yoshizawa, T. & Katayama, T. (1990). Comparative Study On The Natural Occurrence Of Fusarium Mycotoxins (Trichothecenes And Zearalenone) In Corn And Wheat From High- And Low-Risk Areas For Human Esophageal Cancer In China. *Appl Environ Microbiol,* Dec. 56: 3723-3726.

Lutsky, I. & Mor, N. (1981). Experimental Alimentary Toxic Aleukia In Cats. *Lab Anim Sci,* Feb. 31: 43-47.

Ma, J., Ren, Z., Ma, Y., Xu, L., Zhao, Y., Zheng, C., Fang, Y., Xue, T., Sun, B. & Xiao, W. (2009). Targeted Knockdown Of Egr-1 Inhibits Il-8 Production And Il-8-Mediated Invasion Of Prostate Cancer Cells Through Suppressing Egr-1/Nf-Kappab Synergy. *J Biol Chem,* Dec 11. 284: 34600-34606.

Mahmoudi, M. & Gershwin, M.E. (2000). Sick Building Syndrome. Iii. Stachybotrys Chartarum. *J Asthma,* Apr. 37: 191-198.

Maresca, M., Mahfoud, R., Garmy, N. & Fantini, J. (2002). The Mycotoxin Deoxynivalenol Affects Nutrient Absorption In Human Intestinal Epithelial Cells. *J Nutr,* Sep. 132: 2723-2731.

Mccusker, K.P. & Fujimori, D.G. (2012). The Chemistry Of Peptidyltransferase Center-Targeted Antibiotics: Enzymatic Resistance And Approaches To Countering Resistance. *Acs Chem Biol,* Jan 20. 7: 64-72.

Meakins, J.L., Pietsch, J.B., Bubenick, O., Kelly, R., Rode, H., Gordon, J. & Maclean, L.D. (1977). Delayed Hypersensitivity: Indicator Of Acquired Failure Of Host Defenses In Sepsis And Trauma. *Ann Surg,* Sep. 186: 241-250.

Mendoza, M.C., Er, E.E., Zhang, W., Ballif, B.A., Elliott, H.L., Danuser, G. & Blenis, J. (2011). Erk-Mapk Drives Lamellipodia Protrusion By Activating The Wave2 Regulatory Complex. *Mol Cell,* Mar 18. 41: 661-671.

Milbrandt, J. (1987). A Nerve Growth Factor-Induced Gene Encodes A Possible Transcriptional Regulatory Factor. *Science,* Nov 6. 238: 797-799.

Minervini, F., Fornelli, F., Lucivero, G., Romano, C. & Visconti, A. (2005). T-2 Toxin Immunotoxicity On Human B And T Lymphoid Cell Lines. *Toxicology,* May 15. 210: 81-91.

Montana, E., Etzel, R.A., Allan, T., Horgan, T.E. & Dearborn, D.G. (1997). Environmental Risk Factors Associated With Pediatric Idiopathic Pulmonary Hemorrhage And Hemosiderosis In A Cleveland Community. *Pediatrics,* Jan. 99: E5.

Moon, Y., Yang, H. & Lee, S.H. (2007). Modulation Of Early Growth Response Gene 1 And Interleukin-8 Expression By Ribotoxin Deoxynivalenol (Vomitoxin) Via Erk1/2 In Human Epithelial Intestine 407 Cells. *Biochem Biophys Res Commun,* Oct 19. 362: 256-262.

Moon, Y., Yang, H. & Park, S.H. (2008). Hypo-Responsiveness Of Interleukin-8 Production In Human Embryonic Epithelial Intestine 407 Cells Independent Of Nf-Kappab Pathway: New Lessons From Endotoxin And Ribotoxic Deoxynivalenol. *Toxicol Appl Pharmacol,* Aug 15. 231: 94-102.

Nallagatla, S.R., Toroney, R. & Bevilacqua, P.C. (2011). Regulation Of Innate Immunity Through Rna Structure And The Protein Kinase Pkr. *Curr Opin Struct Biol,* Feb. 21: 119-127.

Ng, T.B., Wong, J.H. & Wang, H. (2010). Recent Progress In Research On Ribosome Inactivating Proteins. *Curr Protein Pept Sci,* Feb 1. 11: 37-53.

Novotny, W.E. & Dixit, A. (2000). Pulmonary Hemorrhage In An Infant Following 2 Weeks Of Fungal Exposure. *Arch Pediatr Adolesc Med,* Mar. 154: 271-275.

Nusuetrong, P., Yoshida, M., Tanitsu, M.A., Kikuchi, H., Mizugaki, M., Shimazu, K., Pengsuparp, T., Meksuriyen, D., Oshima, Y. & Nakahata, N. (2005). Involvement Of Reactive Oxygen Species And Stress-Activated Mapks In Satratoxin H-Induced Apoptosis. *Eur J Pharmacol,* Jan 10. 507: 239-246.

Oberholzer, C., Oberholzer, A., Clare-Salzler, M. & Moldawer, L.L. (2001). Apoptosis In Sepsis: A New Target For Therapeutic Exploration. *Faseb J,* Apr. 15: 879-892.

Opal, S.M. (1998). The Uncertain Value Of The Definition For Sirs. Systemic Inflammatory Response Syndrome. *Chest,* Jun. 113: 1442-1443.

Ouyang, Y.L., Azcona-Olivera, J.I., Murtha, J. & Pestka, J.J. (1996). Vomitoxin-Mediated Il-2, Il-4, And Il-5 Superinduction In Murine Cd4+ T Cells Stimulated With Phorbol Ester And Calcium Ionophore: Relation To Kinetics Of Proliferation. *Toxicol Appl Pharmacol,* Jun. 138: 324-334.

Park, S.H., Choi, H.J., Do, K.H., Yang, H., Kim, J. & Moon, Y. (2012). Chronic Nod2 Stimulation Potentiates Activating Transcription Factor 3 And Paradoxical Superinduction Of Epithelial Proinflammatory Chemokines By Mucoactive Ribotoxic Stressors Via Rna-Binding Protein Human Antigen R. *Toxicol Sci,* Jan. 125: 116-125.

Park, S.H., Choi, H.J., Yang, H., Do, K.H., Kim, J. & Moon, Y. (2010). Repression Of Peroxisome Proliferator-Activated Receptor Gamma By Mucosal Ribotoxic Insult-Activated Ccaat/Enhancer-Binding Protein Homologous Protein. *J Immunol,* Nov 1. 185: 5522-5530.

Pestka, J.J. & Smolinski, A.T. (2005). Deoxynivalenol: Toxicology And Potential Effects On Humans. *J Toxicol Environ Health B Crit Rev,* Jan-Feb. 8: 39-69.

Pestka, J.J., Uzarski, R.L. & Islam, Z. (2005). Induction Of Apoptosis And Cytokine Production In The Jurkat Human T Cells By Deoxynivalenol: Role Of Mitogen-Activated Protein Kinases And Comparison To Other 8-Ketotrichothecenes. *Toxicology,* Jan 15. 206: 207-219.

Pestka, J.J., Yan, D. & King, L.E. (1994). Flow Cytometric Analysis Of The Effects Of In Vitro Exposure To Vomitoxin (Deoxynivalenol) On Apoptosis In Murine T, B And Iga+ Cells. *Food Chem Toxicol,* Dec. 32: 1125-1136.

Pestka, J.J., Zhou, H.R., Moon, Y. & Chung, Y.J. (2004). Cellular And Molecular Mechanisms For Immune Modulation By Deoxynivalenol And Other Trichothecenes: Unraveling A Paradox. *Toxicol Lett,* Oct 10. 153: 61-73.

Pieckova, E. & Jesenska, Z. (1999). Microscopic Fungi In Dwellings And Their Health Implications In Humans. *Ann Agric Environ Med,* 6: 1-11.

Puyana, J.C., Pellegrini, J.D., De, A.K., Kodys, K., Silva, W.E. & Miller, C.L. (1998). Both T-Helper-1- And T-Helper-2-Type Lymphokines Are Depressed In Posttrauma Anergy. *J Trauma,* Jun. 44: 1037-1045; Discussion 1045-1036.

Ralph, P. & Nakoinz, I. (1973). Inhibitory Effects Of Lectins And Lymphocyte Mitogens On Murine Lymphomas And Myelomas. *J Natl Cancer Inst,* Sep. 51: 883-890.

Sakiri, R., Ramegowda, B. & Tesh, V.L. (1998). Shiga Toxin Type 1 Activates Tumor Necrosis Factor-Alpha Gene Transcription And Nuclear Translocation Of The Transcriptional Activators Nuclear Factor-Kappab And Activator Protein-1. *Blood,* Jul 15. 92: 558-566.

Sauter, K.A., Melton-Celsa, A.R., Larkin, K., Troxell, M.L., O'brien, A.D. & Magun, B.E. (2008). Mouse Model Of Hemolytic-Uremic Syndrome Caused By Endotoxin-Free Shiga Toxin 2 (Stx2) And Protection From Lethal Outcome By Anti-Stx2 Antibody. *Infect Immun,* Oct. 76: 4469-4478.

Sergent, T., Parys, M., Garsou, S., Pussemier, L., Schneider, Y.J. & Larondelle, Y. (2006). Deoxynivalenol Transport Across Human Intestinal Caco-2 Cells And Its Effects On Cellular Metabolism At Realistic Intestinal Concentrations. *Toxicol Lett,* Jul 1. 164: 167-176.

Shaw, P.J., Ganey, P.E. & Roth, R.A. (2009). Trovafloxacin Enhances The Inflammatory Response To A Gram-Negative Or A Gram-Positive Bacterial Stimulus, Resulting In Neutrophil-Dependent Liver Injury In Mice. *J Pharmacol Exp Ther,* Jul. 330: 72-78.

Shifrin, V.I. & Anderson, P. (1999). Trichothecene Mycotoxins Trigger A Ribotoxic Stress Response That Activates C-Jun N-Terminal Kinase And P38 Mitogen-Activated Protein Kinase And Induces Apoptosis. *J Biol Chem,* May 14. 274: 13985-13992.

Shin, S.Y., Kim, J.H., Baker, A., Lim, Y. & Lee, Y.H. (2010). Transcription Factor Egr-1 Is Essential For Maximal Matrix Metalloproteinase-9 Transcription By Tumor Necrosis Factor Alpha. *Mol Cancer Res,* Apr. 8: 507-519.

Sugimoto, H., Kataoka, T., Igarashi, M., Hamada, M., Takeuchi, T. & Nagai, K. (2000). E-73, An Acetoxyl Analogue Of Cycloheximide, Blocks The Tumor Necrosis Factor-Induced Nf-Kappab Signaling Pathway. *Biochem Biophys Res Commun,* Oct 22. 277: 330-333.

Svirshchevskaya, E., Frolova, E., Alekseeva, L., Kotzareva, O. & Kurup, V.P. (2000). Intravenous Injection Of Major And Cryptic Peptide Epitopes Of Ribotoxin, Asp F 1 Inhibits T Cell Response Induced By Crude Aspergillus Fumigatus Antigens In Mice. *Peptides,* Jan. 21: 1-8.

Tai, J.H. & Pestka, J.J. (1988). Synergistic Interaction Between The Trichothecene T-2 Toxin And Salmonella Typhimurium Lipopolysaccharide In C3h/Hen And C3h/Hej Mice. *Toxicol Lett,* Nov. 44: 191-200.

Tamura, S., Hanada, M., Ohnishi, M., Katsura, K., Sasaki, M. & Kobayashi, T. (2002). Regulation Of Stress-Activated Protein Kinase Signaling Pathways By Protein Phosphatases. *Eur J Biochem,* Feb. 269: 1060-1066.

Tripi, P.A., Modlin, S., Sorenson, W.G. & Dearborn, D.G. (2000). Acute Pulmonary Haemorrhage In An Infant During Induction Of General Anaesthesia. *Paediatr Anaesth,* 10: 92-94.

Underwood, S.L., Kemeny, D.M., Lee, T.H., Raeburn, D. & Karlsson, J.A. (1995a). Ige Production, Antigen-Induced Airway Inflammation And Airway Hyperreactivity In The Brown Norway Rat: The Effects Of Ricin. *Immunology,* Jun. 85: 256-261.

Underwood, S.L., Kemeny, D.M., Lee, T.H., Raeburn, D. & Karlsson, J.A. (1995b). Ricin Increases Ige Levels And Airway Inflammation But Not Hyperresponsiveness In The Rat. *Int Arch Allergy Immunol,* May-Jun. 107: 119-121.

Uzarski, R.L., Islam, Z. & Pestka, J.J. (2003). Potentiation Of Trichothecene-Induced Leukocyte Cytotoxicity And Apoptosis By Tnf-Alpha And Fas Activation. *Chem Biol Interact,* Oct 25. 146: 105-119.

Vaki, I., Kranidioti, H., Karagianni, V., Spyridaki, A., Kotsaki, A., Routsi, C. & Giamarellos-Bourboulis, E.J. (2011). An Early Circulating Factor In Severe Sepsis Modulates Apoptosis Of Monocytes And Lymphocytes. *J Leukoc Biol,* Mar. 89: 343-349.

Valles, P.G., Pesle, S., Piovano, L., Davila, E., Peralta, M., Principi, I. & Lo Giudice, P. (2005). Postdiarrheal Shiga Toxin-Mediated Hemolytic Uremic Syndrome Similar To Septic Shock. *Medicina (B Aires),* 65: 395-401.

Van De Walle, J., During, A., Piront, N., Toussaint, O., Schneider, Y.J. & Larondelle, Y. (2010). Physio-Pathological Parameters Affect The Activation Of Inflammatory Pathways By Deoxynivalenol In Caco-2 Cells. *Toxicol In Vitro,* Oct. 24: 1890-1898.

Van De Walle, J., Romier, B., Larondelle, Y. & Schneider, Y.J. (2008). Influence Of Deoxynivalenol On Nf-Kappab Activation And Il-8 Secretion In Human Intestinal Caco-2 Cells. *Toxicol Lett,* Apr 1. 177: 205-214.

Vandenbroucke, V., Croubels, S., Verbrugghe, E., Boyen, F., De Backer, P., Ducatelle, R., Rychlik, I., Haesebrouck, F. & Pasmans, F. (2009). The Mycotoxin Deoxynivalenol Promotes Uptake Of Salmonella Typhimurium In Porcine Macrophages, Associated With Erk1/2 Induced Cytoskeleton Reorganization. *Vet Res,* Nov-Dec. 40: 64.

Venturini, M.C., Quiroga, M.A., Risso, M.A., Lorenzo, C.D., Omata, Y., Venturini, L. & Godoy, H. (1996). Mycotoxin T-2 And Aflatoxin B1 As Immunosuppressors In Mice Chronically Infected With Toxoplasma Gondii. *J Comp Pathol,* Oct. 115: 229-237.

Viana, M.E., Coates, N.H., Gavett, S.H., Selgrade, M.K., Vesper, S.J. & Ward, M.D. (2002). An Extract Of Stachybotrys Chartarum Causes Allergic Asthma-Like Responses In A Balb/C Mouse Model. *Toxicol Sci,* Nov. 70: 98-109.

Vyleta, M.L., Wong, J. & Magun, B.E. (2012). Suppression Of Ribosomal Function Triggers Innate Immune Signaling Through Activation Of The Nlrp3 Inflammasome. *Plos One,* 7: E36044.

Wang, W.H., Gregori, G., Hullinger, R.L. & Andrisani, O.M. (2004). Sustained Activation Of P38 Mitogen-Activated Protein Kinase And C-Jun N-Terminal Kinase Pathways By Hepatitis B Virus X Protein Mediates Apoptosis Via Induction Of Fas/Fasl And Tumor Necrosis Factor (Tnf) Receptor 1/Tnf-Alpha Expression. *Mol Cell Biol,* Dec. 24: 10352-10365.

Wheeler, D.S. (2009). Death To Sepsis: Targeting Apoptosis Pathways In Sepsis. *Crit Care,* 13: 1010.

Wilhelmsen, C.L. & Pitt, M.L. (1996). Lesions Of Acute Inhaled Lethal Ricin Intoxication In Rhesus Monkeys. *Vet Pathol,* May. 33: 296-302.

Williams, J.M., Lote, C.J., Thewles, A., Wood, J.A., Howie, A.J., Williams, M.C., Adu, D.A. & Taylor, C.M. (2000). Role Of Nitric Oxide In A Toxin-Induced Model Of Haemolytic Uraemic Syndrome. *Pediatr Nephrol,* Oct. 14: 1066-1070.

Wilson, D.N. (2011). On The Specificity Of Antibiotics Targeting The Large Ribosomal Subunit. *Ann N Y Acad Sci,* Dec. 1241: 1-16.

Wong, J., Korcheva, V., Jacoby, D.B. & Magun, B. (2007a). Intrapulmonary Delivery Of Ricin At High Dosage Triggers A Systemic Inflammatory Response And Glomerular Damage. *Am J Pathol,* May. 170: 1497-1510.

Wong, J., Korcheva, V., Jacoby, D.B. & Magun, B.E. (2007b). Proinflammatory Responses Of Human Airway Cells To Ricin Involve Stress-Activated Protein Kinases And Nf-Kappab. *Am J Physiol Lung Cell Mol Physiol,* Dec. 293: L1385-1394.

Wong, S., Schwartz, R.C. & Pestka, J.J. (2001). Superinduction Of Tnf-Alpha And Il-6 In Macrophages By Vomitoxin (Deoxynivalenol) Modulated By Mrna Stabilization. *Toxicology,* Mar 21. 161: 139-149.

Wong, S.S., Zhou, H.R., Marin-Martinez, M.L., Brooks, K. & Pestka, J.J. (1998). Modulation Of Il-1beta, Il-6 And Tnf-Alpha Secretion And Mrna Expression By The Trichothecene Vomitoxin In The Raw 264.7 Murine Macrophage Cell Line. *Food Chem Toxicol,* May. 36: 409-419.

Xu, Y.X., Ayala, A. & Chaudry, I.H. (1998). Prolonged Immunodepression After Trauma And Hemorrhagic Shock. *J Trauma,* Feb. 44: 335-341.

Yan, S.F., Zou, Y.S., Gao, Y., Zhai, C., Mackman, N., Lee, S.L., Milbrandt, J., Pinsky, D., Kisiel, W. & Stern, D. (1998). Tissue Factor Transcription Driven By Egr-1 Is A Critical Mechanism Of Murine Pulmonary Fibrin Deposition In Hypoxia. *Proc Natl Acad Sci U S A,* Jul 7. 95: 8298-8303.

Yang, G.H., Jarvis, B.B., Chung, Y.J. & Pestka, J.J. (2000). Apoptosis Induction By The Satratoxins And Other Trichothecene Mycotoxins: Relationship To Erk, P38 Mapk, And Sapk/Jnk Activation. *Toxicol Appl Pharmacol,* Apr 15. 164: 149-160.

Zhang, P., Langland, J.O., Jacobs, B.L. & Samuel, C.E. (2009). Protein Kinase Pkr-Dependent Activation Of Mitogen-Activated Protein Kinases Occurs Through Mitochondrial Adapter Ips-1 And Is Antagonized By Vaccinia Virus E3l. *J Virol,* Jun. 83: 5718-5725.

Zhou, H.R., Harkema, J.R., Hotchkiss, J.A., Yan, D., Roth, R.A. & Pestka, J.J. (2000). Lipopolysaccharide And The Trichothecene Vomitoxin (Deoxynivalenol) Synergistically Induce Apoptosis In Murine Lymphoid Organs. *Toxicol Sci,* Feb. 53: 253-263.

Zhou, H.R., Harkema, J.R., Yan, D. & Pestka, J.J. (1999). Amplified Proinflammatory Cytokine Expression And Toxicity In Mice Coexposed To Lipopolysaccharide And The Trichothecene Vomitoxin (Deoxynivalenol). *J Toxicol Environ Health A,* May 28. 57: 115-136.

Zhou, H.R., Islam, Z. & Pestka, J.J. (2005). Induction Of Competing Apoptotic And Survival Signaling Pathways In The Macrophage By The Ribotoxic Trichothecene Deoxynivalenol. *Toxicol Sci,* Sep. 87: 113-122.

In: Ribosomes
Editors: Zhou Lin and Wang Liu

ISBN: 978-1-62417-698-2

Chapter 3

The Role of Ribosomes in Mediating Hypoxia Response and Tolerance in Eukaryotes

***Ajit N. Shah*[*], *Md. Maksudul Alam*[*], *Thai Cao and Li Zhang*[†]**
Department of Molecular and Cell Biology,
University of Texas at Dallas, Richardson, TX, US

Abstract

Ribosome biogenesis and protein synthesis are highly energy-consuming processes and therefore, key control points for cellular growth and division under various conditions of stress. Hypoxia is a state of poor oxygen supply that is experienced by diverse organisms under numerous physiological and disease conditions. Hypoxia greatly reduces cellular energy production and hence, organisms must develop effective strategies to cope with this low oxygen state. One common strategy that eukaryotes, ranging from yeast to mammals, employ to adapt under hypoxia is the conservation of cellular energy via suppression of metabolism and protein synthesis. These biological changes occur as a result of coordinated cellular responses orchestrated by various signal transducers and regulators, such as HIFs in higher eukaryotes and Mga2 and Hap1 in yeast. Studies in various eukaryotes have indicated that ribosomes have an essential role in mediating responses to hypoxia. In yeast the expression of various ribosomal and related proteins are altered under hypoxic conditions. Interestingly, it appears that deletion of certain genes involved in ribosome biogenesis, including *NSR1, BUD21, RAI1* or *RPL20A*, minimizes such alterations and enhances the survival of yeast under hypoxic conditions. Such deletions also diminish the effect of key transcriptional regulators, like Swi2 and Swi3, whose targets are altered under hypoxia in the wild-type strain. Similarly, in *C.elegans* reduced activity of arginyl-transfer RNA (t-RNA) synthetase confers hypoxia tolerance. Mammals acquire hypoxia tolerance via global reduction in RNA

---

[*] These authors contributed to this work equally.
[†*] Corresponding author: Li Zhang, Department of Molecular and Cell Biology, The University of Texas at Dallas, Mail Stop RL11, 800 W. Campbell Road, Richardson, TX 75080. Email:li.zhang@utdallas.edu. Phone: 972-883-5757; Fax: 972-883-5759.

translation and ribosome biogenesis. In this chapter, we will review the role of ribosomes in conferring tolerance to hypoxia and discuss future directions to elucidate the molecular mechanisms underlying hypoxia tolerance.

Introduction

Eukaryotic ribosome comprises four different rRNAs and at least 79 proteins; hence, the process of ribosome biogenesis is a highly energy-consuming process [95]. For example, in yeast *S. cerevisiae*, the synthesis of ribosome requires the coordinated action of all three different RNA polymerase enzymes, involves transcription of over 100 different genes, and thereby represents a major fraction of the total biosynthetic activity of the cell [95]. Protein synthesis is another major energy-consuming process. Thus, these two processes are often among the most highly regulated processes for energy conservation under stress conditions. In fact, organisms always adjust their protein biosynthetic capacity in response to nutrient availability, as well as to the presence of environmental stress and other stimuli. This requires coordinated regulation of translation and abundance of ribosomes.

One of the common environmental stresses that eukaryotes experience under various physiological and pathological conditions is the lack of oxygen availability, known as "hypoxia". Hypoxia, by definition, is a pathological condition in which the body as a whole (generalized hypoxia) or a region of the body (tissue hypoxia) is deprived of adequate oxygen supply. Oxygen is essential for the maintenance of life, since it is used as a terminal electron acceptor in oxidative phosphorylation to produce cellular energy. Under normal oxygen tension, eukaryotes oxidize glucose through glycolysis, the tricarboxylic acid (TCA) cycle and oxidative phosphorylation in the mitochondria, thereby maximizing ATP production. Oxidation of one molecule of glucose through glycolysis, TCA cycle and oxidative phosphorylation together yields 38 ATP molecules while consuming 6 molecules of oxygen. However, when eukaryotes encounter hypoxic conditions, they change their physiology into a 'hypoxic mode.' One of the major changes in the cells is their metabolic state; under hypoxia, eukaryotes rely solely on the glycolysis pathway for energy production since oxidative phosphorylation requires oxygen. The hypoxic mode of energy production is inefficient, yielding only 2 ATP molecules from one glucose molecule. In order to survive such energy deficient conditions, organisms need to conserve energy. As major consumers of the cell's energy resources, protein synthesis and ribosome biogenesis is often organism's first choice for reducing energy expenditure. In this review, we will discuss the various aspects of hypoxia response, emphasizing its effect on ribosome biogenesis and protein synthesis.

Hypoxia Is the Hallmark of Various Diseases

Under normal physiological conditions, the human body maintains a steady concentration of oxygen in different tissues. Oxygen is delivered to various tissues of the body by erythrocytes through a network of blood vessels. This steady supply of oxygen ensures proper functioning of the body. Disruption of this system leads to the damage caused by various pathophysiological conditions such as myocardial disease, obstructive lung disease, cancer, cerebral ischemia and preconditioning [79]. Under these conditions, the body mounts various mechanisms to regain the oxygen homeostasis.

Hypoxia is a major feature of solid tumors, which increases treatment resistance and favors cancer progression. Oxygen concentration differs in different areas of the tumor based on the size and vascular network. When the solid tumor grows in size, the rate of proliferation exceeds the rate of vasculature formation. As a result, some regions of the tumor, particularly the center, becomes hypoxic. Tumor hypoxia causes expression and activation of many transcription factors, such as hypoxia inducible factor (HIF) and vascular endothelial growth factor (VEGF). The expression of these factors promotes angiogenesis, which plays an important role in metastasis and continuous proliferation of the solid tumor [80]. The increased expression of HIF-1α, the major regulator of hypoxia, has been observed in various solid tumors. In breast and brain cancers, HIF-1α overexpression has been shown to enhance tumor growth and promotes vascularization [79, 33]. Numerous studies performed over the past decade have shown that hypoxia is an important indicator of the response of the tumor to chemotherapy and radiotherapy [85, 90]. The hypoxic region of the tumor is usually resistant to the radiotherapy. Targeting tumor hypoxia for cancer therapy is an emerging field in cancer research. Studies in mouse xenograft models have demonstrated that topotecan is an effective anti-tumor drug that inhibits HIF-1α activity [75]. Similar studies have been done using tirapazamine, which is activated under hypoxic conditions and is effective in killing radiation-resistant hypoxic cells in solid tumors [33].

Atherosclerosis is responsible for a number of deaths in modern society; it results from the constriction of coronary and cerebral arteries, which in effect leads to an inadequate supply of oxygen to the heart and brain, resulting in ischemia. Ischemia triggers signals that restore the blood flow to the organs by forming new blood vessels to the organ. This response is diminished in diabetic and in elderly people, which results in myocardial ischemia [11, 12, 91]. Myocardial ischemia is a potent inducer of HIF-1 α and VEGF, which are responsible for increased vascularization [57].

Eukaryotes Have Developed Sophisticated Responses to Cope with Low Oxygen Availability at Both the Cellular and Organismal Level

Terrestrial eukaryotes from yeast to mammals often experience episodes of hypoxia; thus, they have developed diverse mechanisms to adapt and survive under hypoxia. For example, western painted turtles that live in North American regions endure low oxygen conditions for months in the winter [10]. Fish such as crucian carp and goldfish can tolerate hypoxia for periods up to 4 months in cold climates [67]. Terrestrial mammals like humans and birds experience hypoxia at high altitude regions. Diving mammals like sperm whale and seals experience hypoxia when submerged in water during a dive. For example, sperm whale can survive submerged in water for 2 hours and seals for a maximum of 1 hour [29]. Yeast such as *S. cerevisiae* also experience hypoxia during fermentation. Others organisms like *C.elegans* and *D. melanogaster* with certain mutations have been isolated that can survive under hypoxia for varying periods of time [3, 81]. These organisms achieve hypoxia tolerance to varying degrees through orchestrated cellular responses focused on reducing the requirement of oxygen for cellular process, avoiding cellular damage, and maintaining the integrity of the organs.

Numerous studies have characterized local (organ-specific) and systemic (whole body) responses that allow these organisms to survive under hypoxic conditions, especially in diving mammals and fish. In particular, aquatic animals try to maintain a steady flow of oxygen to various organs. This is achieved by increasing the diffusion capacity of the gills as well as increasing the number of perfused gill lamellae [74, 97]. Fish and diving mammals also enhance the oxygen-carrying capacity of the blood by increasing hemoglobin level; furthermore, they enhance oxygen delivery to the tissues by increasing myoglobin level. Increased oxygen supply and storage can be an effective mechanism to cope with hypoxia for short durations of time [73]. To cope with long-term hypoxia, fish and amphibians have evolved to accumulate high amount of glycogen in their large muscles and liver.

Another strategy adopted for survival under hypoxia is a partial shutdown of various organs to conserve energy while maintaining their function. In addition to the global adaptive response of reducing metabolic rate and ATP demand, evidences obtained from whole-organism studies showed that various classes of organisms have specific cellular adaptive responses to hypoxia. To conserve energy under hypoxia, the highest energy-consuming cellular processes such as (i) protein synthesis, (ii) protein degradation, (iii) Na^+/K^+ pumping, (iv) urea biosynthesis and (v) glucose consumption in liver cells [43] are deliberately targeted to decrease energy usage. For survival under hypoxia, organisms must adjust to the decreased production of ATP. In order to achieve this demand, organisms have to either partially shut down the most energy-consuming cellular processes mentioned above or increase ATP production by switching to anaerobic pathways [9]. For example, under hypoxia, both glucose consumption and urea biosynthesis in liver cells of turtle drops to zero. Also, protein turnover drops by 90% compared to normoxia. The ATP demand of Na^+/K^+ ATPase also drops by 75% [43]. Similarly, crucian carps, which inhabitate shallow lakes and ponds of northern Europe, downregulates protein synthesis by 95% in liver and muscles, where it accounts for major energy consuming processes [67]. This key strategy of hypoxia adaptation by suppression of protein synthesis is also reported in human cancer cell lines (A549 and HeLa) [52].

Eukaryotes Sense Changes in Oxygen Tension by Using Hemoproteins

Oxygen sensing pathways operate in virtually all eukaryotic cells to mediate responses to hypoxia. Changes in oxygen availability stimulate the activity of hypoxia-responsive proteins, which elicit a range of short- and long-term adaptations to enhance oxygen intake, to increase oxygen delivery, and to conserve cellular energy. In fact, the progress of many developmental, physiological, and pathological processes depends on cell's ability to sense and respond to changes in oxygen levels. To respond efficiently to low-oxygen levels, the change in oxygen levels must be sensed. The search for the oxygen sensor(s) is a rich and rapidly growing field; and a wide variety of molecules have been proposed.

Lines of evidence suggest that membrane proteins containing heme groups may act as oxygen sensors. Cell membrane-bound b-cytochrome-coupled proteins, which are evenly distributed from bacteria to mammals, are potential oxygen sensors [14]. In particular, the cytochrome b558/NADPH oxidase complex is thought to be an oxygen sensor in a range of cell types including phagocytes, oxygen-sensing cells of the carotid body, and airway chemoreceptor cells. Prabhakar and Overholt [72] suggested that cytoplasmic redox-sensitive sites such as NADPH oxidases might be involved in sensing cellular oxygen levels. In organisms ranging from fish to mammals, hypoxia interferes with an iron-dependent interaction between the hypoxia responsive transcription factor hypoxia inducible factor (HIF) and its partner VHL [64]. Under hypoxic conditions, VHL fails to interact with HIF, resulting in the stabilization of HIF-1α. Chandel *et al.* [16] have proposed that the mitochondrion may act as the oxygen sensor. They found that hypoxic induction of oxygen-regulated genes were blocked in hepatocyte rho-0 cells, which lack electron transport chain activity. The cytochrome aa3 has also been identified as the mitochondrial oxygen sensor. Under hypoxia, reduction of O_2 to H_2O by cytochrome aa3 (complex IV) is inhibited, causing the release of electrons upstream at complex III and generation of $O_2^{\cdot-}$. This mitochondrial model of cytochrome aa3 as the oxygen sensor is consistent with an increase in cellular reactive oxygen species (ROS) levels in hypoxia [16].

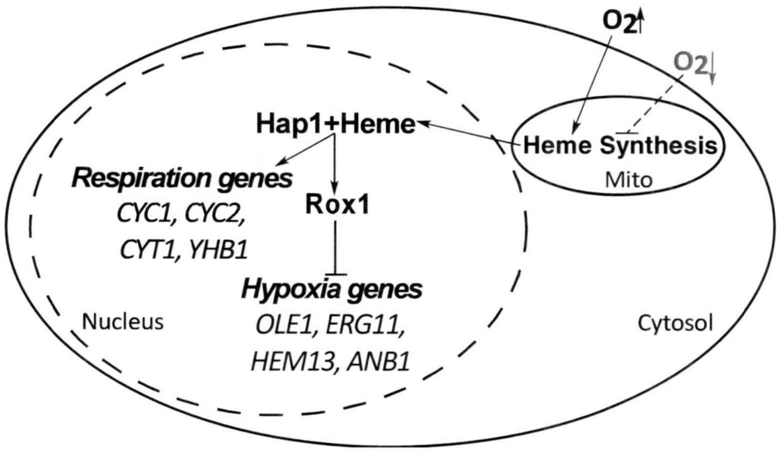

Figure 1. Oxygen sensing in yeast through heme and Hap1interaction. In air, heme is synthesized at a normal level that allows heme binding and activation of Hap1. Activated Hap1induces transcription of Rox1, which inhibits transcription of hypoxia-induced genes. Under hypoxia, heme synthesis is inhibited. As a result, Hap1 inhibits the transcription of Rox1, thereby inducing transcription of hypoxia-induced genes.

In yeast, two well-characterized mechanisms for sensing low oxygen tension have been reported: the first is a specific heme-dependent hypoxia response; the second is a stress response mediated by Rox3 [61]. The heme level in yeast depends on oxygen tension, as oxygen is used as an electron acceptor for heme biosynthesis. Heme is constitutively synthesized at or above oxygen concentrations of 2 μM [103]. However, at levels of oxygen below 2 μM, the process of heme synthesis is tightly coordinated to make the most efficient use of the oxygen-dependent enzymes in heme synthesis [45]. This tight correlation between heme and oxygen tension makes heme a secondary messenger of oxygen level and also helps to mediate the effect of oxygen tension on gene expression. Heme acts as an activator for various transcriptional regulators, including Hap1 (Figure 1). Hap1, previously known as

Cypl, is a transcription factor with a DNA-binding domain at its N-terminus and an activation domain at the C-terminus [103]. Hapl is activated by binding of heme in the presence of adequate oxygen. Once activated, it binds to the transcriptional regulation site of various genes, thereby controlling transcription of various genes involved in oxygen response, such as catalase, cytochromes, hemoglobin and so on [61]. Moreover, activated Hapl also induces the transcription of Rox1, an important transcription factor responsible for the regulation of many hypoxia-induced genes. When the expression of Rox1 is induced by activated Hapl in air, it binds to DNA and represses the expression of genes required for survival under hypoxia. At low oxygen tension, there is not enough heme to activate Hapl, which still binds the promoter of *ROX1* but acts as a transcriptional repressor of *ROX1*. As a result, the expression of *ROX1* is repressed, and genes involved in hypoxia response are transcribed (Figure 1).

Deletion of Genes Involved in Ribosome Biogenesis and Functions Confers Hypoxia Tolerance in Yeast

Performing genome-wide studies has helped discover novel proteins and molecular mechanisms that confer hypoxia tolerance. Many molecular mechanisms are conserved from yeast to humans, and genome-wide experimental resources are readily available in yeast [31, 83]. To date, due to advancements in omics technology, various high throughput libraries of yeast strains are available for genome-wide study. In our recent work, we performed a genome-wide screen to identify genes that influence hypoxia tolerance using yeast knockout library [96]. Of the 5,000 viable yeast strains from the knockout library [96], the initial screening identified 148 strains that showed varying degrees of improved growth under hypoxia versus air compared to the wild-type strain. Of those, 20 showed a substantially improved growth rate under hypoxia. A final quantitative growth rate measurement of the 20 strains helped us identify 5 strains which exhibited substantially improved growth advantage under hypoxia. These strains contain the deletion of *RAI1, NSR1, BUD21, RPL20A*, or *RSM22*. Intriguingly, all of these five genes encode proteins that play a role in ribosomal biogenesis: Rai1 and Nsr1 are required for pre-rRNA processing and ribosomal biogenesis; Bud21 is the component of the small ribosomal subunit (SSU) associated with processing of 35S pre-ribosomal RNA transcript; Rpl20a is a protein component of 60S ribosomal subunit and is a component of mitochondrial small ribosomal subunit [33, 85, 97]. In a similar set of experiments done in *C. elegans*, ethyl methane sulfonate (EMS)-derived mutant worms were screened for hypoxia tolerance [3]. A total of 3884 F1 mutant worm was generated and screened; of these, 14 mutants with hypoxia-resistant phenotypes were recovered. The mutant with the strongest hypoxia tolerance contains mutation in *RRT-1*, which encodes an arginyl-tRNA synthetase. The hypoxia tolerant yeast and worm, as expected, have reduced mRNA translation rate, even in normoxic conditions [3, 83]. In addition, Lori *et al.* showed that the level of hypoxia resistance had a strong inverse correlation with the relative translation rate in worm. Thus, reduction in translation rate is a key feature to confer hypoxia tolerance in eukaryotes. Interestingly, reduced translation has been shown to lengthen the life span of *C. elegans* and other organisms [38, 68, 89]; long-lived mutants were also found to be hypoxia-tolerant [78]. Recent experiments showed that a specific reduction of 60S ribosome subunit levels, due to gene deletion or chemical inhibition, slows aging in yeast [83]. Therefore, reduction of active ribosome slows down translation and increases the hypoxia resistance and life span in eukaryotes.

Deletion of Genes Involved in Ribosome Biogenesis Help Yeast Pre-adopt to Hypoxia and Minimize the Effect of Hypoxia

Molecular characterization of the group of yeast hypoxia-tolerant strains has provided new insights into hypoxia tolerance, specifically the role of ribosomes in hypoxia tolerance [83]. This work showed that hypoxia-tolerant yeast strains minimize the effect of hypoxia on protein synthesis. The wild-type yeast under hypoxia shows a 65% reduction in protein synthesis along with minimum cell proliferation. This reduction in protein synthesis and cell proliferation is an effective mechanism adopted by yeast and other hypoxia-tolerant animals to conserve energy under hypoxia [9, 65, 71]. However, the selected yeast deletion strains Δ*rpl20a* and Δ*bud21* show ~100% increase, Δ*rsm22* and Δ*nsr1* show ~20% increase, and Δ*rai1* shows no difference in the protein synthesis rate under hypoxia (Figure 2) [81]. Notably, the hypoxia-tolerant strains also showed increased cell proliferation under hypoxia, which can be attributed to their unperturbed protein synthesis under hypoxia. Similar results were also reported for hypoxia-tolerant breast cancer cells [36]. Guppy *et al.* found that normal human mammary epithelial (HME) cells show a 62% decrease in protein synthesis under hypoxia along with a 20% decrease in viability, whereas neither viability nor protein synthesis was affected by hypoxia in transformed human mammary epithelial (MCF-7) cells [36]. Thus it appears that the selected hypoxia-tolerant yeast strains and MCF-7 cells are able to proliferate by minimizing the effect of hypoxia on protein synthesis.

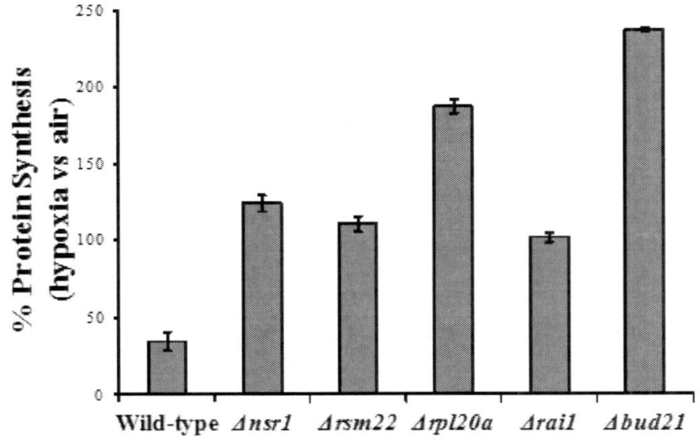

Figure 2. Under hypoxia, hypoxia-tolerant yeast strains exhibit enhanced protein synthesis rate. The protein synthesis is reduced by 65% under hypoxia in the wild-type strain. However, in hypoxia-tolerant deletion strains, the protein synthesis is increased to varying degrees under hypoxia. For experimental details, see [82].

Study of the proteome revealed that the levels of a large number of proteins are altered (upregulated or downregulated) by hypoxia in wild-type yeast are not changed in hypoxia-tolerant strains [83]. Further computational analysis of the proteome changes has helped identify 21 transcriptional regulators which mediate hypoxia-induced changes. These regulators include the known hypoxia–responsive transcription factor Mga2, components of SWI/SNF complex (Snf2, Swi3 and Snf6), and also novel regulators like Aft1, Fhl1, Cst6, and Rap1.

The SWI/SNF complex has previously been reported to be involved in mediating hypoxia response in mammalian cells, where the SWI/SNF complex is known to be essential for HIF activity. Hypoxia-inducible factor-1α (HIF-1α) is a direct target of the SWI/SNF chromatin-remodeling complex, namely BAF57 that binds HIF-1α promoter. The modulation of SWI/SNF levels greatly alters the level of HIF-1α expression and its ability to transactivate target genes [48]. Dastidar *et al.* showed that in wild-type yeast under hypoxia, Swi3 and Snf6 and 4 other SWI/SNF proteins (Snf5, Snf11, Snf12 and Swp82) accumulate in cytosol, and relocalize to nucleus in response to reoxygenation [32, 42]. The redistribution of the SWI/SNF proteins in the cytosol would change the composition of the complex, and presumably affects the function or selectivity of the SWI/SNF complex. This explains why targets of Swi3, Snf2 and Snf6 are altered in wild-type strain under hypoxia. Similarly, other regulators such as Cst6 and Rap1 whose targets are altered by hypoxia in wild-type yeast also accumulate in the cytosol in response to hypoxic conditions [32].

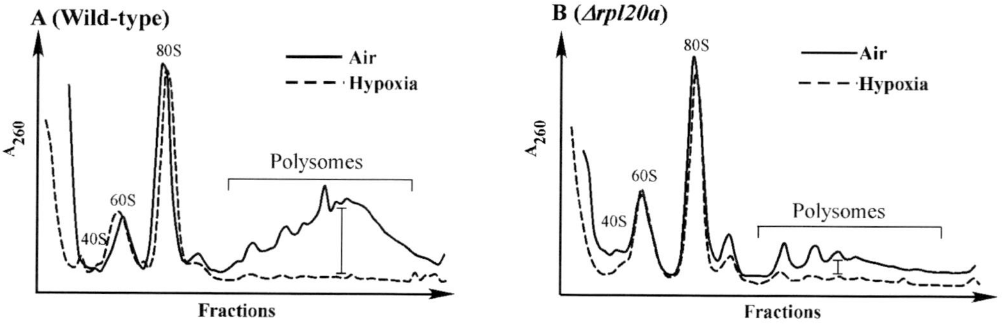

Figure 3. Hypoxia-tolerant yeast strains minimize the effect of hypoxia on polyribosome profile. The amount of polysome is greatly reduced in wild-type yeast. On the other hand, in hypoxia-tolerant deletion strains, the amount of polysome is reduced even in air with almost no further reduction under hypoxia. Consequently, the effect of hypoxia is greatly minimized on polysome profile of the deletion strains.

Interestingly, the targets of the aforementioned regulators are not altered under hypoxia in hypoxia-tolerant strains; therefore, the effects of these regulators are greatly minimized in these knockout strains under hypoxia. As mentioned above, under hypoxia, the wild-type strain show altered levels of the targets of the aforementioned regulators; a similar pattern is observed in air for the deletions strains for these targets. This suggests that the hypoxia-tolerant strains pre-adopt a hypoxia-like state. Similar results were also obtained from the analysis of polyribosome profile (Figure 3). The level of polysomes is greatly reduced in wild-type parent strain under hypoxia, consistent with results of other studies which examine the effect of hypoxia on polyribosome profile [18]. However, the selected deletion strains showed a reduced level of polysomes even in air. Furthermore, the polysome profile of these deletion strains remains unchanged under hypoxia, thereby suggesting that these deletion strains pre-adopt a hypoxia-like polysome profile.

Minimizing Hypoxia-induced Changes on the Expression of Components of Ribosome Increases the Efficiency of Protein Synthesis under Hypoxia

Synthesis of functional ribosomes consumes the most energy in a growing cell under normal conditions. In yeast, there are nearly 2,000,000 ribosomes per cell; with a generation time of ~100 minutes, each yeast cell needs to produce 200 ribosomes per minute. In such rapidly growing yeast cells, the transcription of rRNA genes account for 60% of total transcription. Moreover, the transcription of ribosomal proteins account for 50% of RNA polymerase II transcription and 90% of mRNA splicing [88]. Therefore, coordinated regulation of the ~150 rRNA genes and 137 ribosomal protein (RP) genes is essential to avoid unnecessary use of cell's resources. As the available ATP is greatly diminished in hypoxia, it would be advantageous for the cell to avoid uncoordinated production of different components of ribosomes. In *C. elegans*, deletion of RP genes significantly increases the life span; Kapahi *et al.* reported that, in flies, lifespan is extended by inhibiting S6 kinase (S6K), which, in turn, inhibits translation of RP genes [39]. Life span extension is often associated with increased hypoxia tolerance [78]. The reduction of ribosome biogenesis helps organisms to conserve ATP as ribosome biogenesis is costly in energy. In effect, reduction of ribosome biogenesis slows down the translation rate, which further reduces ATP consumption. Potential advantages of inhibiting ribosome biogenesis may prevent the usage of energy in times of suppressed translation rates. Moreover, the regulation of translation becomes more efficient compared to that achieved by other mechanisms. Together, these factors make organisms better adapt to stress conditions like hypoxia, as well as help them live longer. The proteomic analysis of yeast strains growing under hypoxia has revealed that expression of a significant number of ribosomal proteins have been altered in response to hypoxia, which would make ribosome biosynthesis and function less efficient [83]. The deletion of *NSR1*, *RSM22* or *RPL20A* reduces these changes, thereby minimizing the uncoordinated synthesis of ribosomal components; furthermore, when compared to that of the wild-type strain, the ribosome becomes more efficient in synthesizing proteins, which may account for their enhanced growth under hypoxia [83]. The increased efficiency of ribosome can result from enhanced ribosome recycling/reduced ribosome congestion [29, 55]. Very likely, deletion of one of the five genes help yeast recycle ribosome more efficiently and also reduces ribosome congestion, thereby reversing the deleterious effect of hypoxia on ribosome efficiency.

Diverse Signaling Pathways Can Lead to the Reduction of Translation and Inhibition of Ribosome Biogenesis

Eukaryotes use sophisticated cellular signaling pathways to cope with metabolic and cellular demands under hypoxia [49]. Although different signaling pathways may operate to mediate hypoxia responses in different organisms, the outcomes of hypoxia-induced signal transduction are the same. The two distinctive features of such responses are the suppression of mRNA translation and the inhibition of ribosome biogenesis. In diverse organisms, a common modulator of hypoxia response is the hypoxia inducible factor (HIF) [53, 88, 92]. However, HIF-independent pathways are also involved and are equally important for adaptation under hypoxia. Two important HIF-independent pathways are target of rapamycin (TOR) and unfolded protein response (UPR); both play critical roles in hypoxic adaptations by modulating protein translation and regulating ribosome biogenesis and function [13, 87, 99].

HIF is a heterodimeric complex which is composed of an α and a β subunit [2, 34, 56, 77]. The active form of HIF-αβ complex is known to transactivate a wide variety of genes (1-2% of the genome) by binding to a hypoxia-response element (HRE) in the upstream region of the targets. In particular, the HIF-1 is stabilized under hypoxia and then induces the transcription of many genes that brings about adaptive changes to cope with limited oxygen availability; such genes include erythropoietin (EPO) (induces red blood cell production) [83], vascular endothelial growth factor (VEGF) (promotes angiogenesis) [28], and GLUT1 (increases the efficiency of the glucose uptake) [24]. HIF mRNA contains internal ribosome entry site (IRES); thus, it is not affected by the global suppression of 5'-m7-GTP cap-dependent translation. However, the ribosomal protein S6 (RpS6), modified by various signal transduction pathways, is an important regulator of HIF and VEGF translation; the expression level of these proteins is directly proportional to the phosphorylation of RpS6.

Protein translation is generally inhibited to conserve the cellular energy under severe hypoxic conditions. The target of rapamycin (TOR) complex has emerged as a central regulator of mRNA translation [5, 19, 41, 71]. In mammals, TOR, known as mammalian TOR (mTOR), is a large protein kinase containing sequence similarity to the catalytic domain of the phosphatidiylinositol 3-kinase (PI3K). mTOR causes phosphorylation of eIF4E repressor 4E-BP1 and S6 kinase, thereby inactivating 4E-BP1 and activating S6 kinase (S6K) [70, 99]. The activated S6 kinase stimulates ribosome biogenesis, while inactivated 4E-BP causes cap-dependent translation. mTOR inhibition and subsequent reduction in mRNA translation is necessary to overcome energy stress as the process of mRNA translation is costly in energy. Under other energy stresses such as nutrient deprivation, mTOR activity is inhibited through AMPK and LKB1 [21, 22]. However, under hypoxia, the transcriptional activation of REDD1, a small 25 KD protein with no known functional domain, is responsible for the inhibition of mTOR. It has been proven that REDD1 acts at a point downstream of AKT and AMPK signaling but upstream of TSC2 complex to inhibit mTOR under hypoxia (Figure 4) [99]. In addition, promyelocytic leukemia protein (PML) is also activated in hypoxia that sequesters mTOR into nuclear bodies, thereby inhibiting its activity [84]. The inhibition of TOR activity results in the global decrease of ribosome function through the inhibition of ribosome biogenesis and mRNA translation both in yeast and mammals. As mentioned above, the inhibition of mTOR leads to the hypophosphorylation of RpS6 that causes reduction of the translation of 5'-terminal oligopyrimidine tract (5'TOP) mRNAs, including ribosomal proteins and proteins involved in ribosomal biogenesis.

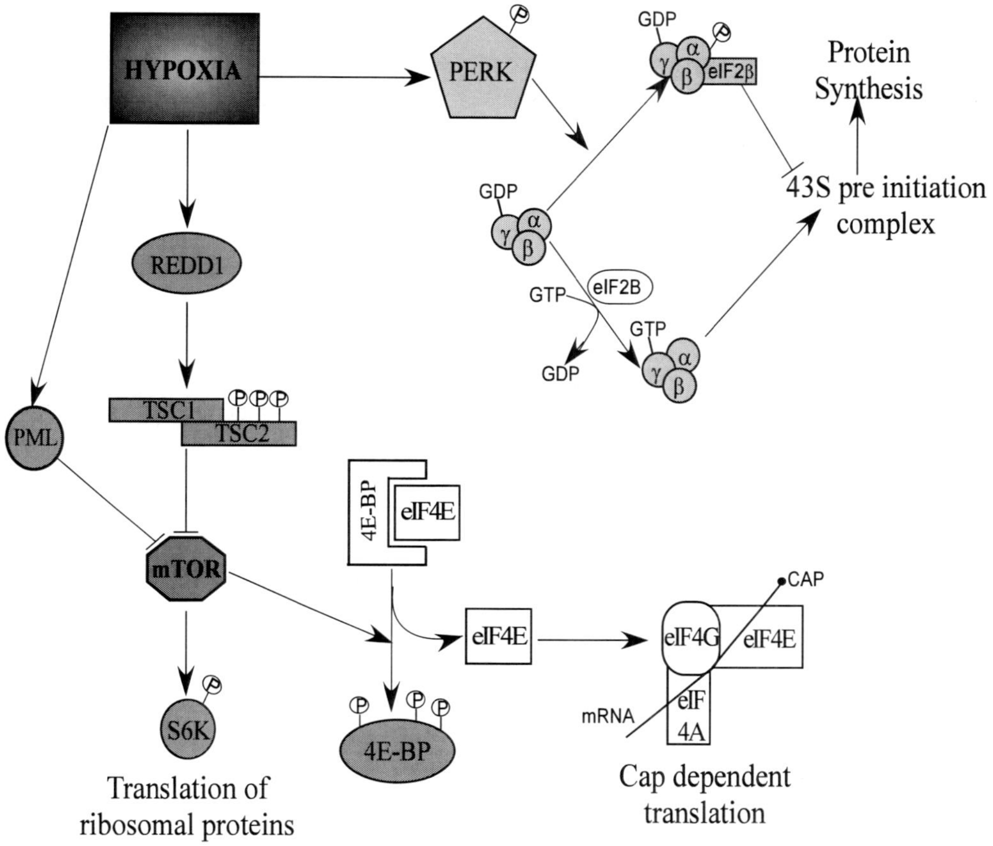

Figure 4. Hypoxia-induced signaling pathways suppress the global and ribosomal protein translation. Hypoxia causes the induction of REDD1, which phosphorylates and activates TSC1/TSC2 complex. The active TSC1/TSC2 complex negatively regulates mTOR, which leads to the hypophosphorylation of 4E-BP and hypophosphorylation of S6K. Consequently, the global and ribosomal protein translation is inhibited. Hypoxia also causes phosphorylation and activation of PERK, which leads to the inhibition of 43S pre-initiation complex, ultimately inhibiting the initiation of translation.

Moreover, the global initiation of mRNA translation is repressed under hypoxia by the rapid phosphorylation of eIF2α, which prevents the formation of an active ternary complex. Formation of the eIF2-GTP/Met-tRNA ternary complex is one of the major control points of translation. At the beginning of protein synthesis, eIF2B catalyzes the hydrolysis of eIF2-bound GTP to GDP; as a result, the initiation factors are released from the ribosome. Therefore, the exchange of eIF2-GDP for GTP is required to start another round of protein synthesis. The formation of eIF2-GTP/Met-tRNA ternary complex is regulated by the phosphorylation of eIF2α, which inhibits the exchange eIF2-bound GTP to GDP without affecting the binding of eIF2B to eIF2α. Under hypoxia, the endoplasmic reticulum kinase PERK is activated by autophosphorylation resulting in phosphorylation of eIF2α as well as inhibition of translation (Figure 4) [5, 52, 58]. However, translation of a relatively small number of factors critical for inducing sustained hypoxia responses, such as VEGF and HIF-1α, is not inhibited [1, 34, 40, 43, 58, 91]. In particular, such genes contain the internal ribosomal entry site (IRES), and therefore do not require eIF4E and the 5'-m7-GTP cap for translation. In addition, UPR induction results in nuclear sequestration of ribosome subunit protein L5, thereby inhibiting ribosome biogenesis and maturation. In this manner, hypoxia-induced signaling pathways minimize ribosome biogenesis in junction with reduction in protein synthesis, as it is unnecessary to synthesize ribosomes at the usual rate while the protein synthesis is being suppressed.

Conclusion

Deletion of genes involved in ribosome biogenesis helps organisms to withstand hypoxic stress. Hypoxia-induced signaling pathways cause the suppression of translation and reduction in ribosome biogenesis. The inhibition of ribosome biogenesis by deleting genes involved in the process facilitates coordination of ribosome production with the rate of mRNA translation, thereby enhancing efficiency in protein synthesis. Consequently, this reduces the unnecessary use of cellular energy for uncoordinated ribosome synthesis. Yeast strains with deletions of ribosomal genes achieve hypoxia tolerance through pre-adopting the hypoxic state by (1) reducing the levels of ribosomes, (2) increasing the efficiency of ribosomes, and (3) minimizing the effect of hypoxia-induced protein changes. Further studies to understand how ribosome efficiency is enhanced in hypoxic yeast strains can provide important new insights into ribosome function and the regulation of ribosome efficiency.

References

[1] Akiri, G., Nahari, D., Finkelstein, Y., Le, S.Y., Elroy-Stein, O., Levi, B.Z. Regulation of vascular endothelial growth factor (VEGF) expression is mediated by internal initiation of translation and alternative initiation of transcription. *Oncogene.* 1998 Jul 16;17(2):227-36.

[2] Amellem, O., Pettersen, E.O. Cell inactivation and cell cycle inhibition as induced by extreme hypoxia: the possible role of cell cycle arrest as a protection against hypoxia-induced lethal damage. *Cell Prolif.* 1991 Mar;24(2):127-41.

[3] Anderson, L.L., Mao, X., Scott, B.A., Crowder, C.M. Survival from hypoxia in C. elegans by inactivation of aminoacyl-tRNA synthetases. *Science.* 2009 Jan 30;323(5914):630-3.

[4] Arany, Z., Huang, L.E., Eckner, R., Bhattacharya, S., Jiang, C., Goldberg, M.A., et al. An essential role for p300/CBP in the cellular response to hypoxia. *Proc Natl Acad Sci U S A.* 1996 Nov 12;93(23):12969-73.

[5] Arsham, A.M., Howell, J.J., Simon, M.C. A novel hypoxia-inducible factor-independent hypoxic response regulating mammalian target of rapamycin and its targets. *J Biol Chem.* 2003 Aug 8;278(32):29655-60.

[6] Bailey, J.R., Val, A.L., Almeida-Val, V.M.F., Driedzic, W.R. Anoxic cardiac performance in Amazonian and north-temperate-zone teleosts. *Canadian Journal of Zoology.* 1999 2012/09/18;77(5):683-9.

[7] Berry, M.J., Knutson, G.S., Lasky, S.R., Munemitsu, S.M., Samuel, C.E. Mechanism of interferon action. Purification and substrate specificities of the double-stranded RNA-dependent protein kinase from untreated and interferon-treated mouse fibroblasts. *J Biol Chem.* 1985 Sep 15;260(20):11240-7.

[8] Bertolotti, A., Zhang, Y., Hendershot, L.M., Harding, H.P., Ron, D. Dynamic interaction of BiP and ER stress transducers in the unfolded-protein response. *Nat Cell Biol.* 2000 Jun;2(6):326-32.

[9] Bickler, P.E., Buck, L.T. Hypoxia tolerance in reptiles, amphibians, and fishes: life with variable oxygen availability. *Annu Rev Physiol.* 2007;69:145-70.

[10] Birse SC, Leonard RB, Coggeshall RE. Neuronal increase in various areas of the nervous system of the guppy, Lebistes. *J Comp Neurol.* 1980 Nov 15;194(2):291-301.

[11] Bosch-Marce, M., Okuyama, H., Wesley, J.B., Sarkar, K., Kimura, H., Liu, Y.V., et al. Effects of aging and hypoxia-inducible factor-1 activity on angiogenic cell mobilization and recovery of perfusion after limb ischemia. *Circ Res.* 2007 Dec 7;101(12):1310-8.

[12] Botusan, I.R., Sunkari, V.G., Savu, O., Catrina, A.I., Grunler, J., Lindberg, S., et al. Stabilization of HIF-1alpha is critical to improve wound healing in diabetic mice. *Proc Natl Acad Sci U S A.* 2008 Dec 9;105(49):19426-31.

[13] Buchberger, A., Bukau, B., Sommer, T. Protein quality control in the cytosol and the endoplasmic reticulum: brothers in arms. *Mol Cell. Oct* 22;40(2):238-52.

[14] Bunn, H.F., Poyton, R.O. Oxygen sensing and molecular adaptation to hypoxia. *Physiol Rev.* 1996 Jul;76(3):839-85.

[15] Carroll, V.A., Ashcroft, M. Targeting the molecular basis for tumour hypoxia. *Expert Rev Mol Med.* 2005 Apr 15;7(6):1-16.

[16] Chandel, N.S., Maltepe, E., Goldwasser, E., Mathieu, C.E., Simon, M.C., Schumacker, P.T. Mitochondrial reactive oxygen species trigger hypoxia-induced transcription. *Proc Natl Acad Sci U S A.* 1998 Sep 29;95(20):11715-20.

[17] Chen, J-J., London, I.M. Regulation of protein synthesis by heme-regulated eIF-2Î± kinase. *Trends in Biochemical Sciences.* 1995;20(3):105-8.

[18] Chen, Y., Feldman, D.E., Deng, C., Brown, J.A., De Giacomo, A.F., Gaw, A.F., et al. Identification of mitogen-activated protein kinase signaling pathways that confer resistance to endoplasmic reticulum stress in Saccharomyces cerevisiae. *Mol Cancer Res.* 2005 Dec;3(12):669-77.

[19] Cherkasova, V.A., Hinnebusch, A.G. Translational control by TOR and TAP42 through dephosphorylation of eIF2alpha kinase GCN2. *Genes Dev.* 2003 Apr 1;17(7):859-72.

[20] Claypool, J.A., French, S.L., Johzuka, K., Eliason, K., Vu, L., Dodd, J.A., et al. Tor pathway regulates Rrn3p-dependent recruitment of yeast RNA polymerase I to the promoter but does not participate in alteration of the number of active genes. *Mol Biol Cell.* 2004 Feb;15(2):946-56.

[21] Clemens, M.J., Bommer, U.A. Translational control: the cancer connection. *Int J Biochem Cell Biol.* 1999 Jan;31(1):1-23.

[22] Corradetti, M.N., Inoki, K., Bardeesy, N., DePinho, R.A., Guan, K.L. Regulation of the TSC pathway by LKB1: evidence of a molecular link between tuberous sclerosis complex and Peutz-Jeghers syndrome. *Genes Dev.* 2004 Jul 1;18(13):1533-8.

[23] Dror, V., Winston, F. The Swi/Snf chromatin remodeling complex is required for ribosomal DNA and telomeric silencing in Saccharomyces cerevisiae. *Mol Cell Biol.* 2004 Sep;24(18):8227-35.

[24] Ebert, B.L., Firth, J.D.,. Ratcliffe, P.J. Hypoxia and mitochondrial inhibitors regulate expression of glucose transporter-1 via distinct Cis-acting sequences. *J Biol Chem.* 1995 Dec 8;270(49):29083-9.

[25] Epstein, A.C., Gleadle, J.M., McNeill, L.A., Hewitson, K.S., O'Rourke, J., Mole, D.R., et al. C. elegans EGL-9 and mammalian homologs define a family of dioxygenases that regulate HIF by prolyl hydroxylation. *Cell.* 2001 Oct 5;107(1):43-54.

[26] Fandrey, J., Gorr, T.A., Gassmann, M. Regulating cellular oxygen sensing by hydroxylation. *Cardiovasc Res.* 2006 Sep 1;71(4):642-51.

[27] Folkman, J. Tumor angiogenesis: therapeutic implications. *N Engl J Med.* 1971 Nov 18;285(21):1182-6.

[28] Forsythe, J.A., Jiang, B.H., Iyer, N.V., Agani, F., Leung, S.W., Koos, R.D., et al. Activation of vascular endothelial growth factor gene transcription by hypoxia-inducible factor 1. *Mol Cell Biol.* 1996 Sep;16(9):4604-13.

[29] Fredrick, K., Ibba, M. How the sequence of a gene can tune its translation. Cell. Apr 16;141(2):227-9.

[30] Gesser, H. The effects of hypoxia and reoxygenation on force development in myocardia of carp and rainbow trout: protective effects of CO2/HCO3. *J Exp Biol.* 1977 Aug;69:199-206.

[31] Ghaemmaghami, S., Huh, W.K., Bower, K., Howson, R.W., Belle, A., Dephoure, N., et al. Global analysis of protein expression in yeast. *Nature.* 2003 Oct 16;425(6959):737-41.

[32] Ghosh Dastidar, R., Hooda, J., Shah, A., Cao, T.M., Henke, R.M., Zhang, L. The nuclear localization of SWI/SNF proteins is subjected to oxygen regulation. *Cell Biosci.* Aug 29;2(1):30.

[33] Giaccia, A.J., Simon, M.C., Johnson, R. The biology of hypoxia: the role of oxygen sensing in development, normal function, and disease. *Genes Dev.* 2004 Sep 15;18(18):2183-94.

[34] Gorlach, A., Camenisch, G., Kvietikova, I., Vogt, L., Wenger, R.H., Gassmann, M. Efficient translation of mouse hypoxia-inducible factor-1alpha under normoxic and hypoxic conditions. *Biochim Biophys Acta.* 2000 Sep 7;1493(1-2):125-34.

[35] Griffith, D.A., Delipala, C., Leadsham, J., Jarvis, S.M., Oesterhelt, D. A novel yeast expression system for the overproduction of quality-controlled membrane proteins. *FEBS Lett.* 2003 Oct 9;553(1-2):45-50.

[36] Guppy, M., Brunner, S., Buchanan, M. Metabolic depression: a response of cancer cells to hypoxia? *Comp Biochem Physiol B Biochem Mol Biol.* 2005 Feb;140(2):233-9.

[37] Hanahan, D., Folkman, J. Patterns and emerging mechanisms of the angiogenic switch during tumorigenesis. *Cell.* 1996 Aug 9;86(3):353-64.

[38] Hand, S.C. Quiescence in Artemia franciscana embryos: reversible arrest of metabolism and gene expression at low oxygen levels. *J Exp Biol.* 1998 Apr;201(Pt 8):1233-42.

[39] Hansen, M., Taubert, S., Crawford, D., Libina, N., Lee, S.J., Kenyon, C. Lifespan extension by conditions that inhibit translation in Caenorhabditis elegans. *Aging Cell.* 2007 Feb;6(1):95-110.

[40] Harding, H.P., Zhang, Y., Ron, D. Protein translation and folding are coupled by an endoplasmic-reticulum-resident kinase. *Nature.* 1999 Jan 21;397(6716):271-4.

[41] Hay, N., Sonenberg, N. Upstream and downstream of mTOR. *Genes Dev.* 2004 Aug 15;18(16):1926-45.

[42] Henke, R.M., Dastidar, R.G., Shah, A., Cadinu, D., Yao, X., Hooda, J., et al. Hypoxia elicits broad and systematic changes in protein subcellular localization. *Am J Physiol Cell Physiol.* 2011 Oct;301(4):C913-28.

[43] Hochachka, P.W., Buck, L.T., Doll, C.J., Land, S.C. Unifying theory of hypoxia tolerance: molecular/metabolic defense and rescue mechanisms for surviving oxygen lack. *Proc Natl Acad Sci U S A.* 1996 Sep 3;93(18):9493-8.

[44] Hochachka, P.W., Lutz, P.L. Mechanism, origin, and evolution of anoxia tolerance in animals. *Comp Biochem Physiol B Biochem Mol Biol.* 2001 Dec;130(4):435-59.

[45] Hon, T., Dodd, A., Dirmeier, R., Gorman, N., Sinclair, P.R., Zhang, L., et al. A mechanism of oxygen sensing in yeast. Multiple oxygen-responsive steps in the heme biosynthetic pathway affect Hap1 activity. *J Biol Chem.* 2003 Dec 12;278(50):50771-80.

[46] Inoki, K., Zhu, T., Guan, K.L. TSC2 mediates cellular energy response to control cell growth and survival. *Cell.* 2003 Nov 26;115(5):577-90.

[47] Ivan, M., Kondo, K., Yang, H., Kim, W., Valiando, J., Ohh, M., et al. HIFalpha targeted for VHL-mediated destruction by proline hydroxylation: implications for O2 sensing. *Science.* 2001 Apr 20;292(5516):464-8.

[48] Jaakkola, P., Mole, D.R., Tian, Y.M., Wilson, M.I., Gielbert, J., Gaskell, S.J., et al. Targeting of HIF-alpha to the von Hippel-Lindau ubiquitylation complex by O2-regulated prolyl hydroxylation. *Science.* 2001 Apr 20;292(5516):468-72.

[49] Jiang, Y., Vasconcelles, M.J., Wretzel, S., Light, A., Gilooly, L., McDaid, K., et al. Mga2p processing by hypoxia and unsaturated fatty acids in Saccharomyces cerevisiae: impact on LORE-dependent gene expression. *Eukaryot Cell.* 2002 Jun;1(3):481-90.

[50] Kenneth, N.S., Mudie, S., van Uden, P., Rocha, S. SWI/SNF regulates the cellular response to hypoxia. *J Biol Chem.* 2009 Feb 13;284(7):4123-31.

[51] Kimata, Y., Ishiwata-Kimata, Y., Yamada, S., Kohno, K. Yeast unfolded protein response pathway regulates expression of genes for anti-oxidative stress and for cell surface proteins. *Genes Cells.* 2006 Jan;11(1):59-69.

[52] Koumenis, C., Naczki, C., Koritzinsky, M., Rastani, S., Diehl, A., Sonenberg, N., et al. Regulation of protein synthesis by hypoxia via activation of the endoplasmic reticulum kinase PERK and phosphorylation of the translation initiation factor eIF2alpha. *Mol Cell Biol.* 2002 Nov;22(21):7405-16.

[53] Kraggerud, S.M., Sandvik, J.A., Pettersen, E.O. Regulation of protein synthesis in human cells exposed to extreme hypoxia. *Anticancer Res.* 1995 May-Jun;15(3):683-6.

[54] Kundaje, A., Xin, X., Lan, C., Lianoglou, S., Zhou, M., Zhang, L., et al. A predictive model of the oxygen and heme regulatory network in yeast. *PLoS Comput Biol.* 2008 Nov;4(11):e1000224.

[55] Kurata, S., Nielsen, K.H., Mitchell, S.F., Lorsch, J.R., Kaji, A., Kaji, H. Ribosome recycling step in yeast cytoplasmic protein synthesis is catalyzed by eEF3 and ATP. *Proc Natl Acad Sci U S A. Jun* 15;107(24):10854-9.

[56] Ladoux, A., Frelin, C. Hypoxia is a strong inducer of vascular endothelial growth factor mRNA expression in the heart. *Biochem Biophys Res Commun.* 1993 Sep 15;195(2):1005-10.

[57] Lando, D., Peet, D.J., Whelan, D.A., Gorman, J.J., Whitelaw, M.L. Asparagine hydroxylation of the HIF transactivation domain a hypoxic switch. *Science.* 2002 Feb 1;295(5556):858-61.

[58] Lang, K.J., Kappel, A., Goodall, G.J. Hypoxia-inducible factor-1alpha mRNA contains an internal ribosome entry site that allows efficient translation during normoxia and hypoxia. *Mol Biol Cell.* 2002 May;13(5):1792-801.

[59] Lee, S.H., Wolf, P.L., Escudero, R., Deutsch, R., Jamieson, S.W., Thistlethwaite, P.A. Early expression of angiogenesis factors in acute myocardial ischemia and infarction. *N Engl J Med.* 2000 Mar 2;342(9):626-33.

[60] Liu, L., Simon, M.C. Regulation of transcription and translation by hypoxia. *Cancer Biol Ther.* 2004 Jun;3(6):492-7.

[61] Lowry, C.V., Zitomer, R.S. ROX1 encodes a heme-induced repression factor regulating ANB1 and CYC7 of Saccharomyces cerevisiae. *Mol Cell Biol.* 1988 Nov;8(11):4651-8.

[62] Mahon, P.C., Hirota, K., Semenza, G.L. FIH-1: a novel protein that interacts with HIF-1alpha and VHL to mediate repression of HIF-1 transcriptional activity. *Genes Dev.* 2001 Oct 15;15(20):2675-86.

[63] Martin, D.E., Soulard, A., Hall, M.N. TOR regulates ribosomal protein gene expression via PKA and the Forkhead transcription factor FHL1. *Cell.* 2004 Dec 29;119(7):969-79.

[64] Maxwell, P.H., Wiesener, M.S., Chang, G.W., Clifford, S.C., Vaux, E.C., Cockman, M.E., et al. The tumour suppressor protein VHL targets hypoxia-inducible factors for oxygen-dependent proteolysis. *Nature.* 1999 May 20;399(6733):271-5.

[65] Mayer, C., Zhao, J., Yuan, X., Grummt, I. mTOR-dependent activation of the transcription factor TIF-IA links rRNA synthesis to nutrient availability. *Genes Dev.* 2004 Feb 15;18(4):423-34.

[66] Mori, K. Signalling pathways in the unfolded protein response: development from yeast to mammals. *J Biochem.* 2009 Dec;146(6):743-50.

[67] Nilsson, G.E., Renshaw, G.M. Hypoxic survival strategies in two fishes: extreme anoxia tolerance in the North European crucian carp and natural hypoxic preconditioning in a coral-reef shark. *J Exp Biol.* 2004 Aug;207(Pt 18):3131-9.

[68] Pan, K.Z., Palter, J.E., Rogers, A.N., Olsen, A., Chen, D., Lithgow, G.J., et al. Inhibition of mRNA translation extends lifespan in Caenorhabditis elegans. *Aging Cell.* 2007 Feb;6(1):111-9.

[69] Patil, C.K., Li, H., Walter, P.. Gcn4p and novel upstream activating sequences regulate targets of the unfolded protein response. *PLoS Biol.* 2004 Aug;2(8):E246.

[70] Pause, A., Belsham, G.J., Gingras, A.C., Donze, O., Lin, T.A., Lawrence, J.C., Jr., et al. Insulin-dependent stimulation of protein synthesis by phosphorylation of a regulator of 5'-cap function. *Nature.* 1994 Oct 27;371(6500):762-7.

[71] Poulin, F., Gingras, A.C., Olsen, H., Chevalier, S., Sonenberg, N. 4E-BP3, a new member of the eukaryotic initiation factor 4E-binding protein family. *J Biol Chem.* 1998 May 29;273(22):14002-7.

[72] Prabhakar, N.R., Overholt, J.L. Cellular mechanisms of oxygen sensing at the carotid body: heme proteins and ion channels. *Respiration Physiology.* 2000;122(2â€"3):209-21.

[73] Ramirez, J.M., Folkow, L.P., Blix, A.S. Hypoxia tolerance in mammals and birds: from the wilderness to the clinic. *Annu Rev Physiol.* 2007;69:113-43.

[74] Randall, D., Daxboeck, C., Randall, W.S., Ha.D.J. Oxygen and Carbon Dioxide Transfer Across Fish Gills. *Fish Physiology: Academic Press;* 1984. p. 263-314.

[75] Rapisarda, A., Uranchimeg, B.,. Sordet, O., Pommier, Y., Shoemaker, R.H., Melillo, G. Topoisomerase I-mediated inhibition of hypoxia-inducible factor 1: mechanism and therapeutic implications. *Cancer Res.* 2004 Feb 15;64(4):1475-82.

[76] Ross, P.L., Huang, Y.N., Marchese, J.N., Williamson, B., Parker, K., Hattan, S., et al. Multiplexed protein quantitation in Saccharomyces cerevisiae using amine-reactive isobaric tagging reagents. *Mol Cell Proteomics.* 2004 Dec;3(12):1154-69.

[77] Schmaltz, C., Hardenbergh, P.H., Wells, A., Fisher, D.E. Regulation of proliferation-survival decisions during tumor cell hypoxia. *Mol Cell Biol.* 1998 May;18(5):2845-54.

[78] Scott, B.A., Avidan, M.S., Crowder, C.M. Regulation of hypoxic death in C. elegans by the insulin/IGF receptor homolog DAF-2. *Science.* 2002 Jun 28;296(5577):2388-91.

[79] Semenza, G.L. Hypoxia-inducible factor 1: oxygen homeostasis and disease pathophysiology. *Trends Mol Med.* 2001 Aug;7(8):345-50.

[80] Semenza, G.L. Oxygen sensing, homeostasis, and disease. *N Engl J Med.* 2011 Aug 11;365(6):537-47.

[81] Semenza, G.L., Roth, P.H., Fang, H.M., Wang, G.L. Transcriptional regulation of genes encoding glycolytic enzymes by hypoxia-inducible factor 1. *J Biol Chem.* 1994 Sep 23;269(38):23757-63.

[82] Semenza, G.L., Wang, G.L. A nuclear factor induced by hypoxia via de novo protein synthesis binds to the human erythropoietin gene enhancer at a site required for transcriptional activation. *Mol Cell Biol.* 1992 Dec;12(12):5447-54.

[83] Shah, A.N., Cadinu, D., Henke, R.M., Xin, X., Dastidar, R.G., Zhang, L. Deletion of a subgroup of ribosome-related genes minimizes hypoxia-induced changes and confers hypoxia tolerance. *Physiol Genomics.* Jul 27;43(14):855-72.

[84] Shaw, R.J., Bardeesy, N., Manning, B.D., Lopez, L., Kosmatka, M., DePinho, R.A., et al. The LKB1 tumor suppressor negatively regulates mTOR signaling. *Cancer Cell.* 2004 Jul;6(1):91-9.

[85] Shinohara, E.T., Maity, A. Increasing sensitivity to radiotherapy and chemotherapy by using novel biological agents that alter the tumor microenvironment. *Curr Mol Med.* 2009 Dec;9(9):1034-45.

[86] Shweiki, D., Itin, A., Soffer, D., Keshet, E. Vascular endothelial growth factor induced by hypoxia may mediate hypoxia-initiated angiogenesis. *Nature.* 1992 Oct 29;359(6398):843-5.

[87] Spriggs, K.A., Bushell, M., Willis, A.E. Translational regulation of gene expression during conditions of cell stress. *Mol Cell. Oct* 22;40(2):228-37.

[88] Sydorskyy, Y., Dilworth, D.J., Yi, E.C., Goodlett, D.R., Wozniak, R.W., Aitchison, J.D. Intersection of the Kap123p-mediated nuclear import and ribosome export pathways. *Mol Cell Biol.* 2003 Mar;23(6):2042-54.

[89] Syntichaki, P., Troulinaki, K., Tavernarakis, N. eIF4E function in somatic cells modulates ageing in Caenorhabditis elegans. *Nature.* 2007 Feb 22;445(7130):922-6.

[90] Tatum, J.L., Kelloff, G.J., Gillies, R.J., Arbeit, J.M., Brown, J.M., Chao, K.S., et al. Hypoxia: importance in tumor biology, noninvasive measurement by imaging, and value of its measurement in the management of cancer therapy. *Int J Radiat Biol.* 2006 Oct;82(10):699-757.

[91] Taylor, C.T., Pouyssegur, J. Oxygen, hypoxia, and stress. *Ann N Y Acad Sci.* 2007 Oct;1113:87-94.

[92] Thangarajah, H., Vial, I.N., Grogan, R.H., Yao, D., Shi, Y., Januszyk, M., et al. HIF-1alpha dysfunction in diabetes. *Cell Cycle.* Jan 1;9(1):75-9.

[93] Tinton, S.A., Buc-Calderon, P.M. Hypoxia increases the association of 4E-binding protein 1 with the initiation factor 4E in isolated rat hepatocytes. *FEBS Lett.* 1999 Mar 5;446(1):55-9.

[94] Travers, K.J., Patil, C.K., Wodicka, L., Lockhart, D.J., Weissman, J.S., Walter, P. Functional and genomic analyses reveal an essential coordination between the unfolded protein response and ER-associated degradation. *Cell.* 2000 Apr 28;101(3):249-58.

[95] Warner, J.R. The economics of ribosome biosynthesis in yeast. *Trends Biochem Sci.* 1999 Nov;24(11):437-40.

[96] Winzeler, E.A., Shoemaker, D.D., Astromoff, A., Liang, H., Anderson, K., Andre, B., et al. Functional characterization of the S. cerevisiae genome by gene deletion and parallel analysis. *Science.* 1999 Aug 6;285(5429):901-6.

[97] Woo, N.Y.S., Wu, R.S.S. Changes in biochemical composition in the red grouper, Epinephelus akaara (Temminck and Schlegel), and the black sea bream, Mylio macrocephalus (Basilewsky), during hypoxic exposure. *Comparative Biochemistry and Physiology Part A: Physiology.* 1984;77(3):475-82.

[98] Wouters, B.G., Koritzinsky, M. Hypoxia signalling through mTOR and the unfolded protein response in cancer. *Nat Rev Cancer.* 2008 Nov;8(11):851-64.

[99] Wouters, B.G., van den Beucken, T., Magagnin, M.G., Koritzinsky, M., Fels, D., Koumenis, C. Control of the hypoxic response through regulation of mRNA translation. *Semin Cell Dev Biol.* 2005 Aug-Oct;16(4-5):487-501.

[100] Wullschleger, S., Loewith, R., Hall, M.N. TOR signaling in growth and metabolism. *Cell.* 2006 Feb 10;124(3):471-84.

[101] Xue, Y., Bai, X., Lee, I., Kallstrom, G., Ho, J., Brown, J., et al. Saccharomyces cerevisiae RAI1 (YGL246c) is homologous to human DOM3Z and encodes a protein that binds the nuclear exoribonuclease Rat1p. *Mol Cell Biol.* 2000 Jun;20(11):4006-15.

[102] Zabetakis, D. Inheritance of suppressors of the drug sensitivity of a NSR1 deleted yeast strain. *Yeast.* 2000 Sep 15;16(12):1147-59.

[103] Zhang, L. Heme Biology – The secret life of heme in regulating diverse biological processes.: World Scientific.; 2011.

In: Ribosomes
Editors: Zhou Lin and Wang Liu
ISBN: 978-1-62417-698-2

Chapter 4

Putative Protein-encoding Genes within Mitochondrial rDNA and the D-loop Region

Hervé Seligmann*
Department of Ecology, Evolution and Behaviour, The Hebrew University of Jerusalem, Jerusalem, Israel
Department of Life Sciences, Ben Gurion University, Beer Sheva, Israel

Abstract

Analyses suggest that protein coding genes occur within ribosomal RNA genes, enabling to code for additional proteins without increasing genome size. Stop codons can be translated by antisense antitermination (suppressor) tRNAs with anticodons matching stop codons, hence overlap coding might not be limited to stopless 'open reading frames'. Alignment analyses detect several putative protein coding genes within mitochondrial rRNAs and the mitochondrial D-loop region that include stops. These putative overlapping genes are probably functional because corresponding RNA abundances in GenBank's EST database increase with the length of the putative overlap coding region (but not for regions not involved in overlap coding), and predicted protein secondary structures resemble those formed by their presumed homologues from GenBank. In addition, numbers of putative overlap coding genes increase with predicted capacities of antisense tRNAs to match stop codons when comparing mitochondrial genomes from each primate and *Drosophila* species. In human mitochondrial genomes, single nucleotide polymorphisms are less frequent in putative rRNA-encoded overlapping protein coding genes than in the rest of each 12S, 16S rRNAs and the D-loop, as one would expect if the coding function of these regions cause excess purifying selection against polymorphisms. However, further analyses based on polymorphism frequencies and proportions of pathogenic mutations among these yield partially inconclusive results, probably because of weaknesses intrinsic to indirect estimation of purifying selection as indirect evidence for gene function. Previous analyses showed that overlapping protein

* Email: podarcissicula@gmail.com.

coding genes in frameshifted and antisense sequences of regular mitochondrial protein coding genes coevolve with capacities of antisense tRNAs to translate stop codons, reassigning stop codons to code for amino acids: UAR are reassigned to glutamine in primates, tryptophane in turtles, and serine in *Drosophila*; AGR stops are reassigned to arginine in primates and glycine in the olive ridley (*Lepidochelys*) and lysine in other turtles. Analyses of mitochondrial rRNAs converge with analyses of protein coding genes and strengthen the hypothesis that mitochondrial genomes code according to a stopless genetic code for previously undetected, cryptic overlapping protein coding genes. It is probable that pathogenicity of at least some polymorphisms is due to the previously unsuspected function of rRNA and D-loop regions as protein coding regions.

Introduction

Open reading frames occur within rRNA (as introns in nuclei, organelles and bacteria, see Hausner et al 1999 and Johansen et al 2007; but also as overlapping genes in nuclei, (Kermekchiev and Ivanova 2001; Coelho et al 2002; Bonawitz et al 2008; Galopier and Hermann-Le Denmat 2011)), enabling to code for additional proteins without increasing genome size. However, it seems that not only stopless 'open reading frames' may code for proteins: the predicted occurrence of antisense antitermination (suppressor) tRNAs matching stop codons (Seligmann 2010) suggests that even regions that include stops may be translated and function as overlapping genes. Indeed, previous analyses showed that overlapping protein coding genes within frameshifted and antisense sequences of regular mitochondrial protein coding genes coevolve with the capacity of antisense tRNAs to translate stop codons (Seligmann 2011a, 2012a; Faure et al 2011). In relation to stop codons, these analyses suggest that overlap coding and suppressor tRNA properties are optimized for requirements of efficient translation. Alignment analyses indicate that mitochondrial stop codons are reassigned to code for amino acids: UAR stops are putatively reassigned to glutamine in primates (Seligmann 2011a), tryptophane in turtles (Seligmann 2011b), and serine in *Drosophila* (Seligmann 2012a); AGR stops are reassigned to arginine in primates (Seligmann 2011a; Faure et al 2011), glycine in the olive ridley (*Lepidochelys olivacea*) and lysine in other turtles (Seligmann 2011b). Additional analyses show that predicted overlapping protein coding genes are more conserved than regular main frame protein coding sequences that code only for one protein: their nucleotide contents do not match predictions made by replicational mutation gradients (A->G and C->T) that characterize (main frame) third codon positions when these are not involved in overlap coding, and that reflect gradients in singlestrandness during replication by the heavy strand (Seligmann 2012b). Frequencies of these substitutions increase with the duration of singlestrandedness during replication (i.e. Krishnan et al 2004).

Here I explore whether mitochondrial DNA sequences templating for 12S and 16S rRNAs include similar protein coding genes that reassign an amino acid to stops, using methods similar to those used previously for protein coding genes (Seligmann 2011a, 2012a; Faure et al 2011). Both DNA strands are examined as it is known that each mitochondrial DNA strand is transcribed in its entirety (Tabak et al 1983; Clayton 1984). Analyses include the mitochondrial control region (D-loop) of mtDNA which contains the heavy strand origin of replication and has been shown to be the most variable region in both vertebrates and invertebrates. It is not known to contain protein coding genes.

Results

Putative Protein Coding Genes

I extracted the human 12S, 16S and D-loop sequences (NC 012920) and translated with 'transeq' (http://www.ebi.ac.uk/Tools/emboss/transeq/index. html) the 6 potential frames (3 for each positive, negative DNA strands), inserting asterisks (*) at stop codons. For *Homo sapiens*, this yields 17 candidate protein sequences, which were detected by BLAST (Altschul et al 1997, Altschul et al 2005), GenBank's online software that searches for matches between a focal sequence and sequences existing in GenBank. I used standard default search parameters.

Table 1. Overlapping protein coding regions predicted by GenBank's BLAST alignment analyses with proteins in GenBank for human mitochondrial 12s, 16s rRNAs and the Dloop

Gene	Fr	Residues	N	%	Align with	Stops
12S	3	71-99	28	73	c4b-binding protein alpha chain, *Macaca mulatta*, XP_001110596	R
12S	3	162-193	31	66	Unnamed prot., *Tetraodon nigroviridis*, CAF92285	
12S	3	184-213	29	75	ND1, *Pongo abelii*, XP_002812499	R
12S	5	123-160	37	87	Hyp. prot., *Plasmodium chabaudi*, CAH88354	
16S	1	251-458	207	52	Hyp. 18K prot. - goldfish mitochondrion, JC1348	10, 1R
16S	1	459-498	39	85	PTH-responsive osteosarcoma D1 prot., *Homo sapiens*, AAD25980	R
16S	1	405-487	60	58	Hyp. prot., *Helicobacter felis* ATCC 49179, CBY83691	6
16S	2	60-121	62	81	Hyp. prot., *Rickettsia rickettsii* str. Hino, AFB28311	7, 5R
16S	2	417-477	61	57	Hyp. prot., *Anaerostipes* sp., EFV21076	2, 1R
16S	3	321-344	23	96	Humanin, *Homo sapiens*, AAK50430	R
16S	3	6-24	18	100	Unknown prot., *Homo sapiens*, AAH00845	
16S	3	271-293	23	83	Unnamed prot., *Macaca fascicularis*, BAE90024	
16S	5	164-216	52	55	Hyp. lipoprot., *Campylobacter jejuni*, EDZ04993	4, 2R
16S	6	2-99	59	57	Cell wall hydrolase, *Rothia dentocariosa*, EFJ76323	9, 1R
Dloop	1	17-97	80	100	CoxII/D-loop DNA fusion protein, *Homo sapiens*	
Dloop	2	17-37	20	100	CoxII/D-loop DNA fusion protein, *Homo sapiens*	
Dloop	2	127-165	48	75	Unnamed prot., *Rattus norvegicus*	
Dloop	3	13-45	32	100	CoxII/D-loop DNA fusion protein, *Homo sapiens*	
Dloop	4	216-258	43	72	Hyp. prot., *Pan troglodytes*	

Columns are: 1. Human sequence identity; 2. Frame that yields alignment (1-3 for sense (positive) strand; 4-6 for antisense (negative) strand; 3. First and last residues in alignment; 4. Alignment length; 5. Percent similarity between aligning proteins; 6. GenBank protein aligning with hypothetical proteins translated from human mitochondrial rRNAs or D-loop; 7. Number of stops, and number aligning with R in alignment.

BLAST analyses yield positive hits for each of the three regions (four for 12S rRNA, ten for 16S rRNA, and five for the D-loop, see Table 1), covering 26.7, 54.2 and 40.9% of these regions, respectively. Hence numerous candidate overlapping protein coding regions occur within each human mitochondrial 12S rRNA, 16S rRNA, and the D-loop.

The spatial distributions of the regions coding for the putative proteins is not random. These regions seem to be clustered. For 12S rRNA, the four proteins from GenBank match sequences presumably coded by two regions on the 12S rRNA, one close to its 5' extremity (from nucleotide 860 to 946 according to GenBank's site numbering), and the three others from 1122 to 1288. This includes two contiguous regions on the positive strand templating for 12S rRNA, and one on the complementary negative strand of that same region. This type of spatial clustering occurs also for candidate protein coding genes in 16S rRNA. Here, nucleotides 1688-1744 close to the 5' extremity of that gene, and nucleotides 1848-2032 presumably code each for a protein. The rest of the candidate protein coding genes are within two regions, nucleotides 2419-2739 and 2881-3162, coded on the positive and the negative strands in these regions. For the D-loop, all candidate genes are within the region of nucleotides 82-540, hence one single cluster that involves both positive and negative DNA strands in the 3' half of that region.

This clustering, for each of the three regions, seems non-random and in itself suggests that the hits from BLAST are not a matter of chance. However, misannotations of rRNAs have reputedly produced artefactual proteins (Tripp et al 2011; Eberhardt et al 2012). It is probable that relatively short alignments, especially when these represent only a fraction of the protein in GenBank, are artefacts. For example, the short alignments with AAH00845 and BAE90024 could result from chance, however note that similarities are 100 and 83%. On the other side of the spectrum, almost the total length of AFB28311 aligns with a sequence translated from 16S rRNA, a probable candidate protein coding gene. Many alignments are with proteins from primates or vertebrates, notably humanin, a nucleus-encoded gene with mitochondrial origins (Tajima et al 2002), a putative 18K protein from goldfish mitochondria (Peng et al 1992). Other alignments are usually with bacteria that infect humans (EFV21076, EFJ76323) and/or are proteobacteria (CBY83691, AFB28311, EDZ04993), the likely ancestors of mitochondria. Most of the more likely cases, considering protein origins, alignment quality and length are from 16S rRNA sequences. Hence overlap coding in 16S rRNA seems the most likely at this stage.

The following analyses intend to test whether overlap coding is probable when considering various properties of the predicted alignments.

Predicted Protein Secondary Structures

Comparisons between secondary structure predictions for the aligning protein sequences can indicate whether the alignments in Table 1 reflect putative functional proteins, and functional protein coding genes. I used the online available software Concord (http://helios.princeton.edu/CONCORD/, Wei et al 2011) to predict peptide secondary structures. Its output indicates the broad types of secondary structures in which each residue is involved. I used it to predict secondary structures formed by amino acid sequences translated from mitochondrial rRNAs and the D-loop and aligning with GenBank proteins, and predicted secondary structures for these amino acid sequences from GenBank. Then predicted secondary structure categories of putatively homologous residues were compared.

Predicted secondary structures were identical for putatively homologous residues in 79% of the cases, averaging this percentage across all putative proteins in Table 1 (excluding those, for the D-loop, where alignments had 100% identities). This is to be compared with the similarities (also in Table 1) between the linear sequences, which are on average 66%. Hence on average, the putative homologous amino acid sequences form more similar secondary structures than one would expect considering similarities between their linear amino acid sequences.

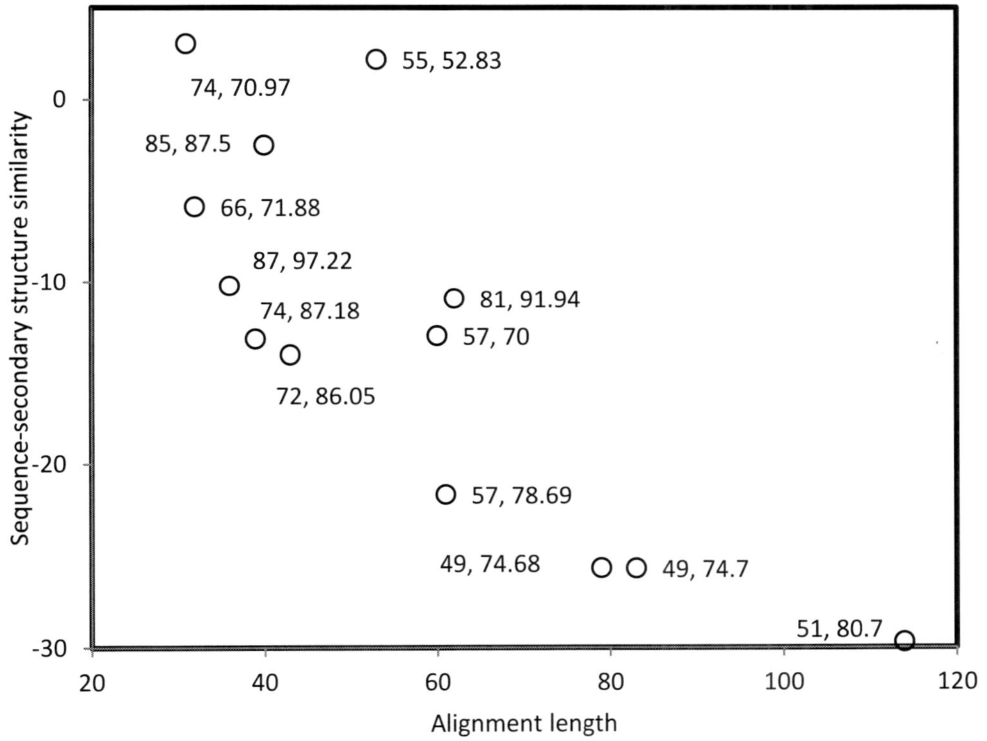

Figure 1. Subtraction of similarity between predicted secondary structure similarities for aligning sequences in Table 1 (only for sequences >29 residues) from similarities between linear sequences in alignments in Table 1. Numbers near datapoints indicate similarities between sequences followed by similarities between predicted secondary structures. The longer the alignment, the more similar the predicted secondary structures as compared to similarities between sequences.

This qualitative difference was obtained for all but two alignments in Table 1 (the unnamed protein from *Tetraodon nigroviridis*, in 12S rRNA, and the hypothetical lipoprotein from *Campylobacter jejuni*, in 16S rRNA), where similarity between sequences is greater than the identity between predicted secondary structures. These alignments were among the shortest ones, and indeed the subtraction of secondary structure identity from sequence similarity decreases with the length of the alignment (Figure 1). This suggests that the quality of secondary structure prediction increases with sequence length. Hence the cases where sequence similarity is greater than identities between the secondary structures are probably artefacts due to short sequence length. Even if one includes these short alignments, qualitatively, secondary structure identity was greater than sequence similarity in 85% of the alignments, which is a clear majority according to a one tailed sign test (P = 0.00562). This line of evidence, though relatively preliminary considering the type of analysis, suggests that the sequences translated from mitochondrial rRNAs and the D-loop could represent functional proteins, as these are predicted to form secondary structures that are similar to those formed by the presumed homologous proteins, and this beyond what would be expected considering similarities between their sequences.

RNA Abundances of Putative Protein Coding Genes

Analyses of GenBank's EST (expressed sequence tags) database showed that for putative overlap coding genes within *Drosophila*'s regular mitochondrial main frame protein coding genes, extents of overlap coding correlates positively with GenBank EST abundances, for overlapping genes coded by the negative and positive DNA strands (see figures 11 and 13, respectively, in Seligmann 2012a). Hence the sequence (according to the positive DNA strand) of each of the various regions suspected to code for proteins within each human mitochondrial 12S, 16S rRNAs and D-loop were analysed separately by BLAST in relation to GenBank's EST database, limiting searches to EST sequences from human origin. Similar analyses were done, as control, for sequences within 12S, 16S rRNAs and the D-loop that are not suspected of overlap coding. Here too, search parameters were the standard default ones.

BLAST returns, for each input sequence, the 100 ESTs that yield the highest similarity with the input, indicating whether the hit is with the input sequence (here the 'positive' strand), or with its inverse complementary (antisense) sequence (here the negative strand). Figure 2 plots numbers of EST hits with the input (positive) sequence as a function of the length of the input sequence, distinguishing between regions that are not suspected of overlap protein coding according to Table 1 (hollow circles in Figure 2), those presumably coding only on the sense (positive) DNA strand (filled circles), and those presumably coding on the antisense (negative) DNA strand (triangles). For the sake of clarity of graphical presentation, locations of sequences are indicated only for putative protein coding regions.

For the regions that are not suspected to code for overlapping genes, EST numbers apparently decrease slightly with sequence length ($r = -0.429$, two tailed $P = 0.216$, discontinuous line in Figure 2). This statistically non-significant tendency serves as control for similar analyses for the rest of the data. For regions on the sense (positive) DNA strand and putatively coding for a protein, EST numbers increase with sequence length ($r = 0.8184$, one tailed $P = 0.0065$). This statistically significant increase, contrasted with the opposite tendency for regions not suspected to code for overlapping genes, suggests that overlap coding associates positively with EST abundance. A one tailed test was used because EST abundance is expected to increase with overlap coding. This principle is confirmed when analysing the putative overlap coding genes situated on the antisense (negative) DNA strand. For these, the trend in Figure 2 is expected to be negative, remembering that the y axis in Figure 2 plots the number of ESTs, among 100 hits, with the positive strand. Indeed, EST abundance decreases with sequence length ($r = -0.56$, one tailed $P = 0.07$) for the putative overlap coding sequences located on the antisense sequence. Hence here too, presumed overlap coding increases expression as reflected by EST abundance.

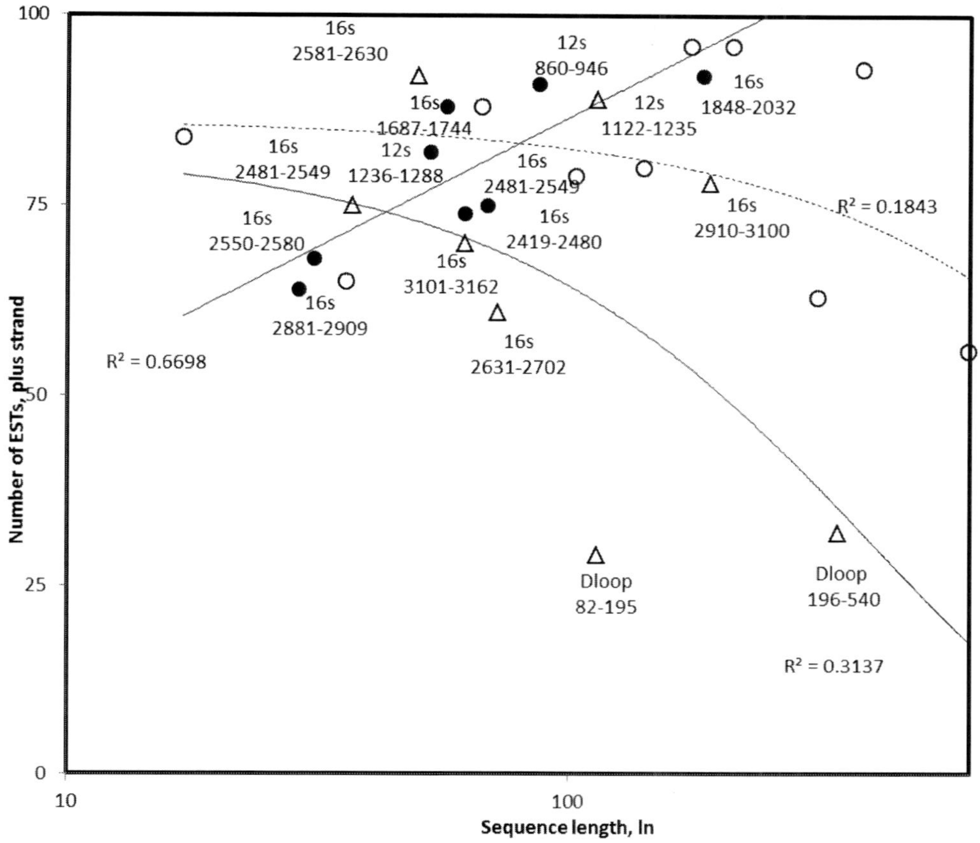

Figure 2. Human EST numbers in GenBank's database matching the positive strand of various regions of human mitochondrial rRNAs and D-loop as a function of the region's length, separating between regions not predicted to code for proteins according to Table 1 (empty circles, discontinous regression line), those expected to code for proteins on the positive (sense) strand (filled circles, increasing continous line) and those expected to code on the negative (antisense) strand (triangles, decreasing continous line). ESTs are detected by BLAST, which returns the 100 transcripts best fitting with the input sequence. The y axis indicates the number of hits with transcripts of the positive strand, to obtain numbers of BLAST hits with the negative strand one has to subtract the value on the y axis from 100.

The combined significance of the tests for positive and negative strands, according to Fisher's method for combining P values that sums -2xLnPi where Pi is the statistical significance of the ith test, with i ranging from 1 to k tests and where this sum is a chisquare statistic with 2xk degrees of freedom yields chisquare = 15.27 and P = 0.0042 (k = 2 in this case), confirming at the level of combined positive and negative DNA strands this positive association between length of presumed overlap coding regions and corresponding EST abundances.

Coevolution between Overlap Coding and Mitochondrial Antisense Suppressor tRNAs

The putative overlap coding regions include numerous stop codons (last column in Table 1). Hence activity by suppressor tRNAs (also termed antitermination tRNAs, Seligmann

2010a) is required for their expression. In order to test this, analyses such as those done for human 12S, 16S rRNAs and the D-loop were repeated for mitochondrial genomes from other primate species, and for all *Drosophila* species for which a complete mitochondrial genome was available in GenBank in summer 2010.

Antisense tRNA anticodon identity and their capacity to form the cloverleaf secondary structure that is essential for translational activity were predicted by tRNAScan-SE (http://lowelab.ucsc.edu/tRNAscan-SE/, Lowe and Eddy (1997), Schattner et al (2005)), using the organellar model for tRNAs and setting the cutoff value for the 'Cove' score to -100 when analysing antisense tRNA sequences. The secondary structure component of Cove was used to estimate the capacity to form cloverleaf secondary structures by antisense tRNA sequences. Note that Cove is a log ratio between a score for the sequence and random sequences. Negative values of Cove indicate a tendency for cloverleaf formation that is lower than random, while positive ones indicate a greater capacity. The number of putative overlapping genes within the three regions for primates was plotted as a function of the mean Cove of antisense tRNAs with anticodons matching UAR stop codons (Figure 3). A similar analysis was done for *Drosophila* species (Figure 4).

These analyses yield for each primates and *Drosophila* the expected positive correlation between numbers of putative overlapping protein coding genes detected by BLAST within 12S and 16S rRNAs and the D-loop (for primates: $r = 0.359$, one tailed $P = 0.0185$; Spearman rank non-parametric correlation coefficient $rs = 0.305$, one tailed $P = 0.025$; for *Drosophila*: $r = 0.415$, one tailed $P = 0.062$, $rs = 0.438$, one tailed $P = 0.051$). This yields a combined P value (according to Fisher's method to combine P values) of $P = 0.0089$ for Pearson parametric correlation coefficients r and $P = 0.0098$ for rs analyses. This result shows that the capacity to translate stops, as estimated by cloverleaf formation by antisense antitermination tRNAs coevolves with the need to translate such stops, as estimated by the number of overlap coding genes predicted by BLAST for mitochondrial rRNAs and D-loops. It is a strong indication that these putative protein coding regions are actually expressed.

The latter result is not only remarkable in relation to expression of previously undetected protein coding genes within mitochondrial rRNAs and the D-loop. This is an independent confirmation, using these three regions, of results obtained for putative overlap coding genes within regular protein coding regions for the same primate and *Drosophila* mitochondrial genomes (Seligmann 2011a, 2012a; Faure et al 2011). Considering that at this stage, indications that proteins corresponding to these genes exist (and were explored) only for one of them (GAU, Faure et al 2011), such computational 'bioinformatic' confirmations are an important step in relation to assess the existence of overlap coding by stopless genetic codes, in addition to the EST abundance analyses presented in the previous section.

Another important aspect in relation to positive coevolution between antisense antitermination tRNAs and overlapping genes in mitochondrial rRNAs and the D-loop is that at this point, the evidence for the existence of antisense tRNAs is based mainly on computations (Seligmann 2010a, 2011c). The only piece of evidence with a 'physical' component is that in *Drosophila*, computational estimates of adaptedness for translation of mitochondrial antisense tRNAs correlate positively with antisense tRNA abundance within GenBank's EST library (figure 10 in Seligmann 2012a). The structural Cove scores used in analyses (Figures 3 and 4) are 'only' computationally predicted. These are not the result of direct measures of cloverleaf formation. However, the fact that the same predicted cloverleaf formation Cove scores coevolves positively within each primates and *Drosophila* (the only groups investigated in this respect), with numbers of overlapping genes predicted for each mitochondrial protein coding genes, and the mitochondrial rRNAs and the D-loop cannot be the result of chance.

This means that the computational predictions probably reflect a biological reality. Assuming translation of overlapping genes by antisense suppressor tRNAs explains parsimoniously and coherently patterns observed between disparate sequence properties and types (alignments with proteins, and tRNA properties). Such a coherent body of positive evidence, strengthened by its repeatability, considering that it is based on independent observations from primates and *Drosophila*, cannot be dismissed on the base of lack of previous evidence. It justifies overlap coding by overlapping stopless genetic codes as a valid working hypothesis, despite its apparent provocative boldness.

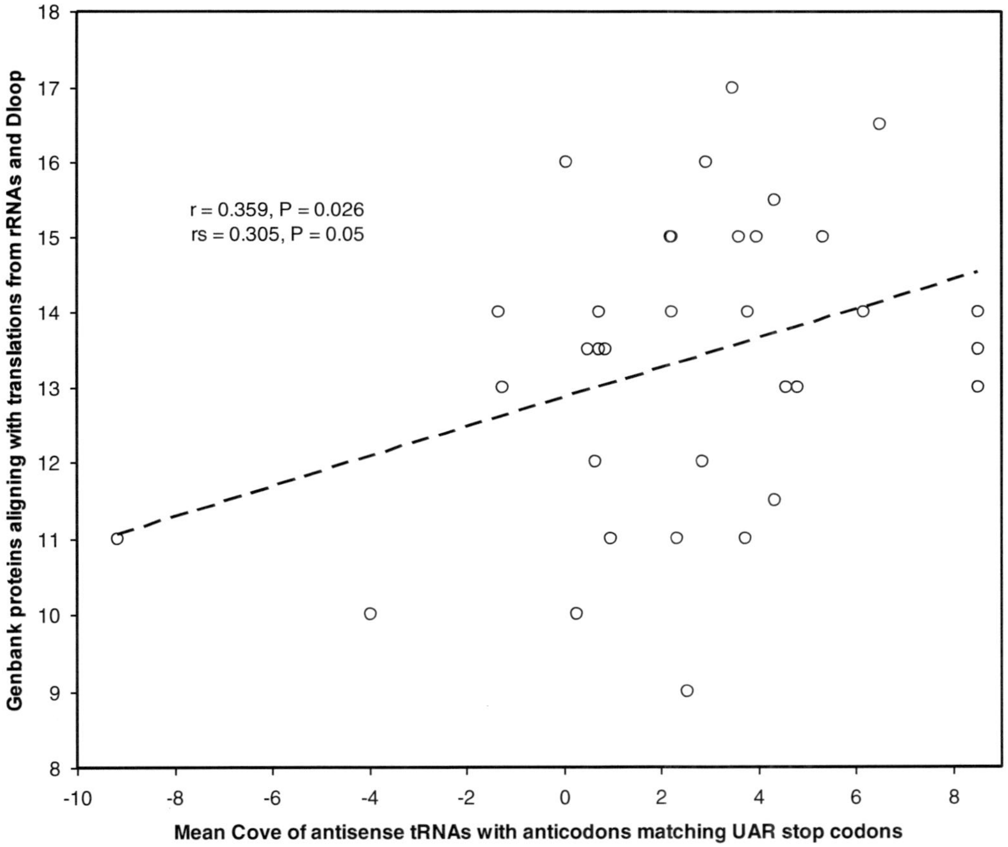

Figure 3. Number of protein sequences in Genbank aligning according to 'BLAST' with protein sequences translated from any frame of primate mitochondrial rRNAs and D-loops, as a function of the average capacity of antisense tRNAs with anticodons matching UAR stop codons in these primate genomes to form cloverleaf secondary structures, as estimated by the structural component of the Cove index from tRNAscan-Se for these antisense tRNA sequences.

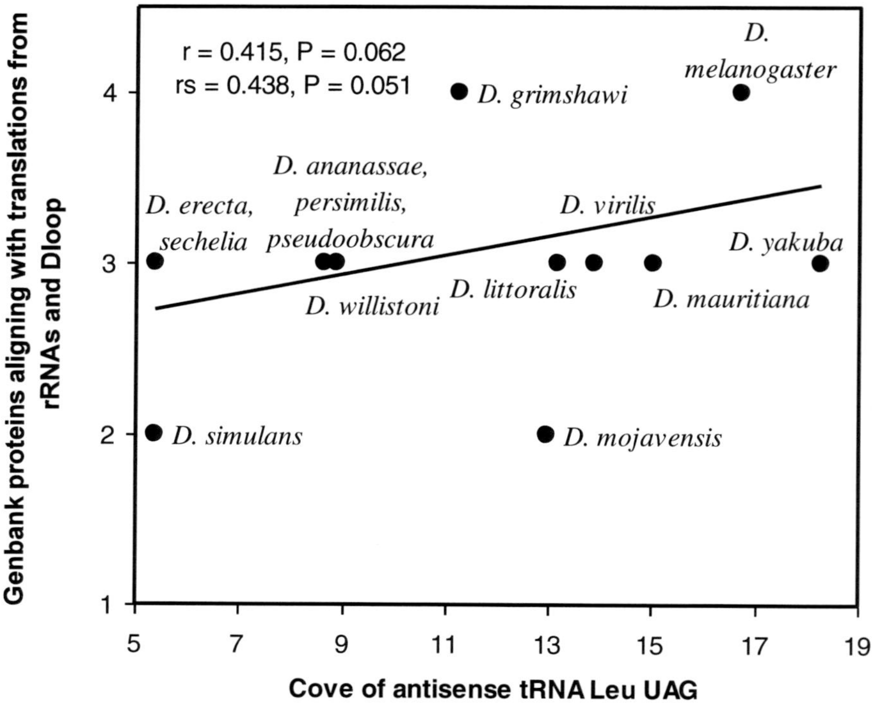

Figure 4. Number of protein sequences in Genbank aligning according to 'BLAST' with protein sequences translated from any frame of *Drosophila* rRNAs and D-loop, as a function of the average capacity of antisense of tRNA Leu CUA with anticodons matching UAG stop codons in these primate genomes to form cloverleaf secondary structures, as estimated by the structural component of the Cove index from tRNAscan-Se for these antisense tRNA sequences.

Single Nucleotide Polymorphisms and Overlap Coding in Mitochondrial rRNAs and the D-loop

The above discussion justifies considering at this point that most of the candidate overlapping protein coding genes are valid protein coding genes. One cannot be sure that each of these sequences is an actual gene, nor whether additional sequences within these regions might function as protein coding genes, however, overall, one can assume that at least most of the candidate genes are real protein coding genes that are occasionally expressed.

This means that the same DNA has more than one function, the one known previously (i.e., templating for rRNAs), and the one of coding for proteins. Consistently with that, it is much more probable that a mutation within these regions disturbs at least one of the functions, hence will be counterselected. This principle of greater purifying selection against mutations within regions with predicted overlap coding functions predicts that one should observe fewer polymorphisms within these regions, as compared to other regions. I used the polymorphism information available for human mitochondrial genomes and compiled at mtDB (http://www.mtdb.igp.uu.se/, Ingman and Gyllensten 2006) and at MITOMAP (http://www. mitomap.org/bin/view.pl/MITOMAP/VariantSubmissionList; and http://www.

mitomap.org/bin/view.pl/MITOMAP/MutationsRNA and http://www.mitomap.org/bin/view.pl/MITOMAP/Mutations CodingControl, Ruiz-Pesini et al 2007) and calculated proportions of polymorphic sites for regions that are not predicted involved in overlap coding for the mitochondrial rRNAs and the D-loop, as compared to those where Table 1 predicts overlap coding for these sequences. Figure 5 plots this proportion as a function of the number of overlap coding frames predicted according to Table 1. The proportion of polymorphic sites indeed decreases, as predicted, with the number of overlap coding frames in a sequence.

Values next to datapoints indicate numbers of sites with single nucleotide polymorphisms, followed by the total number of sites in that category. P values are for two sided Fisher exact tests between that point and the next point on that line, indicating that for each additional overlap coding frame, a significant decrease in the proportion of polymorphic sites occurs.

Figure 5 suggests that the more overlap protein coding frames BLAST predicted for a given sequence, the fewer polymorphic sites in that sequence. This presumably results from the sum of (and interaction between) selection pressures on its protein coding functions. The more protein coding functions, the greater the constraints on mutations, hence such mutations are rarer in the natural *Homo sapiens* population. It is notable that mitochondrial sequences that template for rRNAs, but are not predicted to code for overlapping protein coding genes, are more polymorphic than regions predicted to code for such proteins. This result strengthens the hypothesis of overlap coding by these regions.

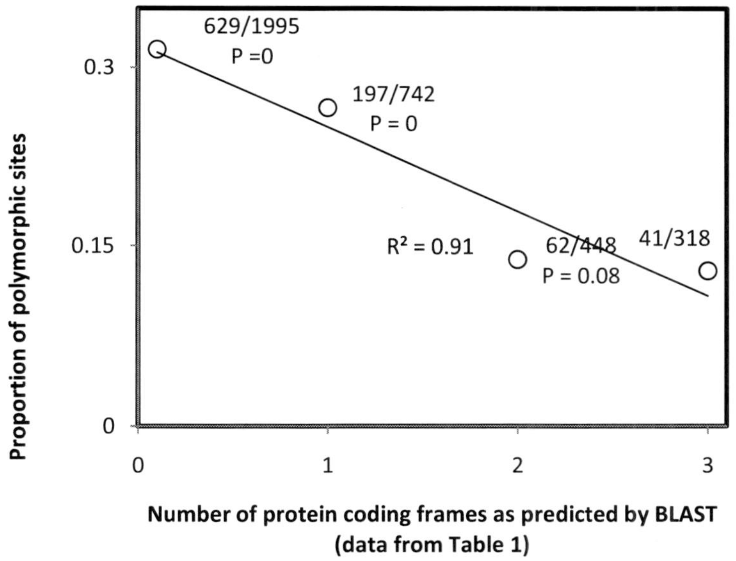

Figure 5. Proportion of polymorphic sites in different regions of human mitochondrial 12S, 16S rRNAs and the D-loop, as a function of numbers of protein coding frames predicted by Table 1.

However, analyses in Figure 5 pool polymorphisms from mitochondrial rRNAs and the D-loop, which have very differing biological functions. Analysing separately each of these regions yields similar results for 12S rRNA (r = -0.974, one tailed P = 0.073), 16S rRNA (r = -0.992, one tailed P = 0.0039), but results are much weaker for the D-loop (r =-0.489, one tailed P = 0.255). The combined P for these three sequences according to Fisher's method for combining Ps yields chisquare = 19.06, which with 2*k=6 degrees of freedoms yield P = 0.004. Results suggest that while overall, principles on polymorphisms are confirmed, in the details, the role of variation within the D-loop is not well understood, suggesting a larger amount of positive selection on D-loop polymorphisms.

Pathogenic Mutations in Overlap Coding Genes

The analyses in the previous section are based on the assumption that most mutations are counterselected. However, the high level of polymorphism observed for the D-loop suggests that in that case, this may not be true. The issue of polymorphism counterselection and the difficulty to interpret data is further indicated by considering pathogenic polymorphisms, as these are indicated at MITOMAP (Ruiz-Pesini et al 2007). For the same reason as one expects less polymorphism with increasing level of sequence functional importance, one expects a greater proportion, among these polymorphisms, of polymorphisms associated with pathologies. However, this prediction is based on a simplistic scenario. This is because mutations might be so detrimental that they are lethal, and hence incompatible with survival, even in a diseased state. Hence such mutations would not be observed. That situation would expect less pathogenic polymorphisms among sites with a high functional importance (because a higher proportion of mutations would be lethal and remain unobserved), and more among those with lower functional importance. Considering this, it is not surprising that proportions of pathogenic mutations do not increase (in a statistically significant way) with the number of coding frames (as used as x axis in Figure 5), for any of the three analysed regions, though qualitatively, the proportion increases (not in a statistically significant manner) as predicted for 16S rRNA and the D-loop. For 12S rRNA, the proportion decreased, which does not fit predictions by the simple scenario that the more functions a sequence has, the more probable that mutations are pathogenic. Hence, though interesting, analyses using proportions of polymorphic sites, and the proportion, among these, of pathogenic polymorphisms, are not strong tests for the functionality of regions. They may, however, indicate what types of functional pressures sequences undergo, and the relative strengths of these pressures. In addition, the pathogenic phenotypes associated with mutations in one or the other region may be indicative of the physiological function of the sequence. The data on pathogenic mutations for the mitochondrial rRNAs and the D-loop do not yield any indications in this respect, and the proportion of these did not differ among the various regions involved, or not involved, in overlap coding. This direction of analyses, however, would be worth exploring further, especially to test for associations with specific pathological phenotypes, and in relation to overlap coding genes detected within regular mitochondrial main frame protein coding genes.

Codon Positions in Putative Overlapping Genes and Polymorphisms

The data on polymorphism frequencies and frequencies of pathogenic polymorphisms can be analysed according to similar principles, but based on a different biological background. In codons, the three nucleotides bear differently on coding properties, with mutations at third codon position being either neutral (they do not alter the coded amino acid) or causing mainly conservative amino acid replacements. At the other extreme, mutations at second codon positions tend to have drastic effects on the coded amino acid, swapping between amino acids with usually very different basic physico-chemical properties. Mutations at first codon positions are intermediate between 3[d] and 2nd codon positions, but tend to resemble more 2nd codon positions.

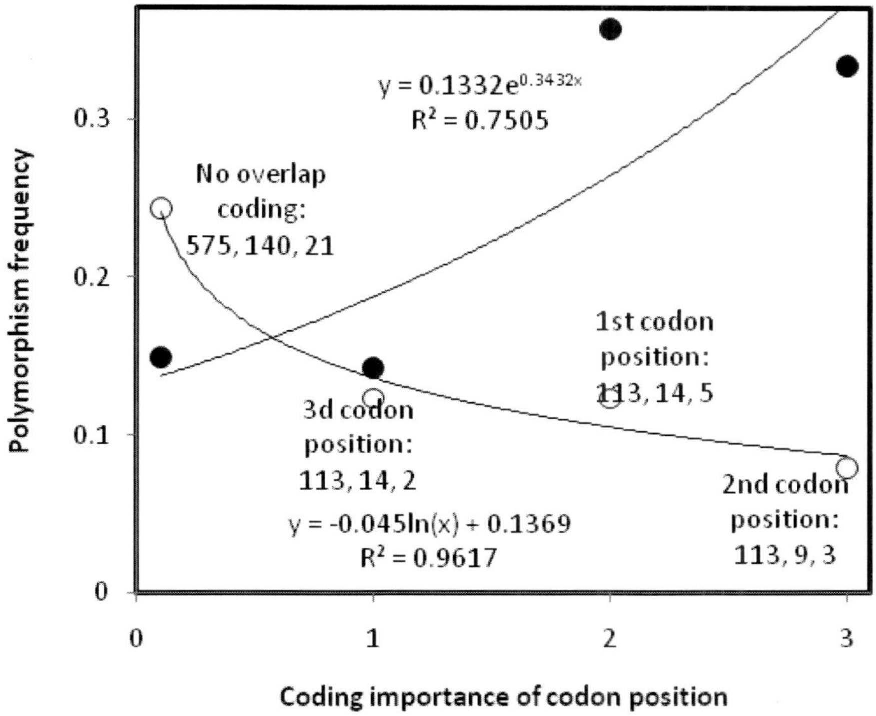

Figure 6. Polymorphism frequency (open circles) and frequency of pathogenic polymorphisms among all polymorphisms (filled symbols) in predicted 16s rRNA overlapping protein coding genes (excluding regions where more than one overlapping gene putatively occurs) as a function of the coding importance of a codon position (mutations at 3[d] codon positions have the least impact on coding and get value '1', '0' is for regions not expected involved in overlap coding). Polymorphisms seem counterselected the greater their putative impact on coding, as expressed by their decreasing frequencies (open symbols) and the increasing proportion, among them, of mutations found associated with diseases (filled symbols). Values near datapoints indicate the total number of sites involved, the total number among them for which polymorphisms occur, and the number of polymorphic sites where pathogenic mutations occur. The decrease in polymorphism frequency between '0' and '1' is significant according to a one tailed Fisher exact test ($P = 0.0027$), but not between subsequent datapoints. The increase in the proprotion of pathogenic mutations between categories '0-1' and '2-3' is also significant according to a one tailed Fisher exact test ($P = 0.026$).

Accordingly, one can expect that for putative overlapping protein coding genes described in Table 1, single nucleotide polymorphisms should be relatively frequent at 3[d] codon positions, and least frequent at 2nd codon positions, with 1st codon positions having intermediate frequencies of polymorphisms. Using a semi-quantitative approach to estimate mutation impacts at each codon position, I assigned 3[d] codon positions 'coding importance' '1', '2' for 1st codon positions, and '3' for 2nd codon positions as done previously for overlapping genes within regular main frame human mitochondrial protein coding genes (Seligmann 2012b). Sequences that are not involved in overlap coding get coding importance '0' in this approach. Polymorphism frequencies and frequencies of pathogenic ones among these are expected to decrease, respectively increase along that gradient, following the simplistic scenario that does not consider mutation lethality for pathogenic mutations. Figure 6 shows this for 16S rRNA regions involved in overlap coding, excluding those where more than one overlapping coding frame is expected to occur (coding importance is more complex to quantify in such regions), however, no clear trend occurs for 12S rRNA and the D-loop (not shown). This reinforces the point that selection on mutations probably follows different scenarios, according to different regions and the sequence's function(s).

This positive result for 16s means that even the fine structure of the overlapping coding regions is confirmed by polymorphism data. The fact that such a confirmation is not obtained for other regions should not be seen as evidence that predicted overlap coding genes within these regions is an artefact, but that solely these are not confirmed by analyses of polymorphisms: lack of evidence is not evidence of lack. This is further discussed below.

Conclusion

Alignment analyses of human mitochondrial 12S and 16S rRNAs and D-loop sequences translated into protein sequences indicate that regions that include stop codons code for previously undetected overlapping protein coding genes. The interpretation that these regions are probably functional protein coding genes is strengthened by a number of independent tests: a) the predicted secondary structures formed by the amino acid sequences translated from the putative mitochondrial overlapping genes in the mitochondrial rRNAs and the D-loop are very similar to those predicted for their putative homologues in GenBank, and this similarity between secondary structures increases with similarities between the aligning linear sequences; b) abundances of RNAs corresponding to putative overlapping genes increase with the length of the predicted overlapping genes, and this is not the case for mitochondrial sequences within these regions that are not predicted involved in overlapping protein coding genes; c) coevolution exists between predicted capacities of mitochondrial antisense tRNAs to suppress stops (stop suppression is necessary to enable read-through of these genes that include stop codons) and the numbers of predicted overlapping protein coding genes in the mitochondrial rRNAs and the D-loop. A similar coevolution exists between these tRNAs and numbers of overlapping genes within regular mitochondrial main frame protein coding genes (Seligmann 2011a, 2012a), the expression of one of the proteins (Gau) coded by these genes has been indicated by a monoclonal antibody (Faure et al 2011).

The secondary structure evidence should be considered as the weakest evidence, as predicted secondary structures of proteins are relatively inaccurate. The EST data is a relatively strong indication as it consists of actual transcripts detected by other studies. Analyzes of polymorphisms can only be considered as confirmative when these yield positive results, but coevolution with antisense antitermination tRNAs is probably strong evidence. In addition, RNA transcripts were detected for two of the predicted genes (from goldfish, Peng et al 1992) and the neuroprotective factor peptide humanin (Tajima et al 2002) which seems encoded by both the nucleus and the mitochondrial genome (Hashimoto et al 2001, Guo et al 2003, Bodzioch et al 2009).

Hence predicted protein secondary structure, RNA transcript abundances, and coevolution between putative coding and the translational apparatus indicate that a large number of mitochondrial proteins are coded by previously undetected mitochondrial genes whose expression is switched 'on' by translational activity of suppressor tRNAs, whether of endogenous mitochondrial origins (Seligmann 2010a) or imported from the cytosol (Schneider 2011).

Analyses of single nucleotide polymorphisms in human mitochondrial rRNAs and the D-loop do not systematically confirm that the putative overlapping protein coding genes are functional. Polymorphism frequencies decrease with the number of putative protein coding frames, as expected if one assumes that mutations are usually counterselected, and this pressure should increase with the number of coding functions assumed by a nucleotide. However, patterns in relation to proportions of pathogenic polymorphisms are not so clearly explained by purifying selection. This may be due to high mutation lethality for regions with great functional importance, as well as to positive selection for some polymorphisms, especially in the Dloop. Analyses of polymorphism frequencies considering the putative positions of polymorphic nucleotide sites within codons of predicted protein coding sequences does yield a clear picture for 16S rRNA (polymorphism frequencies decrease along the gradient of predicted non-coding regions, to 3^{d}, 1st and 2nd codon positions; and the proportions of pathogenic mutations among these polymorphisms increase along that same gradient in coding 'importance'). However, there were no clear tendencies for 12S rRNA and the D-loop.

According to the latter observations on polymorphisms, a conservative interpretation would be to consider that the working hypothesis is upheld for 16S rRNA overlap coding, but not for 12S rRNA and the D-loop. However, in this context, it is important to weigh the different types of evidence, in order to avoid an overconservative stance. The analyses predicting patterns in polymorphism frequencies presumably reflect purifying selection, which supposedly reflects the functional importance of a nucleotide. The estimation of selection intensity is indirect. In addition, selection intensity is an indirect estimate of functionality of a sequence. Hence it should not be surprising that considering this, not all analyses based on this indirect approach yield positive evidence. This has to be contrasted with the analyses that explore more direct functional predictions (coevolution with the translational apparatus, prediction of protein secondary structures, and RNA abundances), and which all yield quite clear, positive evidence. Getting partially inconclusive results from the most indirect method should not invalidate the positive evidence that more direct methods produce, especially that in this case the negative results are a lack of evidence, not evidence of lack. Inconclusive results do not invalidate positive ones based on independent data.

In addition, analyses of polymorphism frequencies in relation to codon positions rely on the assumption that mutations at 3^{d} codon positions and in presumably non-coding regions are relatively neutral. This is not necessarily the case, and numerous examples indicate that mutations that are presumably 'silent' from a coding point of view are nevertheless not neutral, because various properties can affect translation, from RNA secondary structure (Krishnan et al 2008) to codon-anticodon mismatches (Seligmann 2010c; 2011d), to refer only to a few examples.

In addition, it is possible and even plausible that overlapping protein coding genes are located in rRNA (and D-loop) regions that bear relatively low functional constraints from the point of view of their supposed primary function (as rRNAs or D-loops). Hence it is possible that the regions that are not predicted to code for overlapping proteins are under greater (or similar) selection only as the result of their regular primary function as rRNAs or D-loop, even when the other regions cumulate selective pressures from their regular functions with those associated with overlap coding functions. Hence polymorphism frequencies may be lower in these non-coding sequences than in the putative protein coding regions because selection may be weaker on the latter, despite cumulation of several functions on the former.

According to that principle, proportions of pathogenic mutations might be greater in non-overlapping regions than in putatively overlapping protein coding regions. Patterns for analyses of polymorphisms in the D-loop could fit this scenario (not shown). Considering this, negative and inconclusive results from analyses based on estimates of presumed selection intensity and their effects on mutations rates (and the resulting polymorphisms) should not be seen as invalidating other types of positive evidence. This is also valid for methods designed to detect selection on overlapping genes and broadly based on ratios between synonymous versus non-synonymous variation, contrasted between putative overlapping genes and non-overlapping gene regions (i.e. Sabath et al 2008; 2009), which frequently yield negative results in relation to the functionality of at least one of the overlapping genes (Sabath and Graur 2010). While positive results on selection based on such analyses is good evidence for the functionality of overlapping genes, negative results should be considered for what they are: inconclusive, and not as arguments to dismiss positive evidence from different, frequently more direct approaches, that suggest the existence of putative overlapping genes.

A last point to discuss here is the issue of translational termination at conditions where suppressor tRNAs are active in mitochondria. Recent results and analyses suggest that termination of mitochondrial translation is not a simple mechanism: AGR 'stops' cause -1 frameshifting that 'creates' a 'regular' UAG stop (Richter et al 2010; Temperley et al 2010; Young et al 2010; Chrzanowska-Lightlowers et al 2011; Korostelev 2011; Pacheu-Grau et al 2012). I suspect that the relatively strong coevolution observed between stop codons in overlap coding genes and suppressor tRNAs, which is greater than coevolution between other sense and antisense tRNAs and usages of their corresponding codon families (in primates and *Drosophila*, see respectively Seligmann 2010b, 2012a) reflects the necessity for suppressor tRNAs to compete with termination factors, a competition that does not occur for other codons. Hence termination might occur along regular mechanisms, but would be a more probabilistic process than in the total absence of suppressor tRNAs.

The evidence at hand suggests that several previously undetected overlapping protein coding genes occur within mitochondrial 12S and 16S rRNA and D-loop sequences. All analyses confirm this for the largest sequence, 16S rRNA, whose size may enable less structural constraints on each region, allowing to code for additional functions. However, it is very probable that at least some of the candidate overlapping protein coding genes in the 12S rRNA and the D-loop are functional, as most bioinformatics analyses confirm this, and analyses of polymorphisms are simply inconclusive. It seems probable that at conditions where suppressor tRNAs are active in mitochondrial translation, adequate experiments will detect proteins with amino acid sequences matching many among those predicted by the putative overlapping genes and presented here. This means that one has to consider that sequences templating for mitochondrial rRNAs and the D-loop have additional, coding functions, and that pathogenicities of some mutations within these regions are due to the perturbation of these protein coding regions.

Acknowledgments

Comments by the reviewer, Eric Faure, greatly improved this study.

References

Altschul, SF; Madden, TL; Schäffer, AA; Zhang, J; Zhang, Z; Miller, W; Lipman, DJ. Gapped BLAST and PSI-BLAST: a new generation of protein database search programs. *Nucleic Acids Res* 1997, 25, 3389-3402.

Altschul, SF; Wootton, JC; Gertz, EM; Agarwala, R; Morgulis, A; Schäffer, AA; Yu YK. Protein database searches using compositionally adjusted substitution matrices. *FEBS J* 2005, 272, 5101-5109.

Bodzioch, M; Lapicka-Bodzioch, K; Zapala, B; Kamysz, W; Kiec-Wilk, B; Dembinska-Kiec, A. Evidence for potential functionality of nuclearly-encoded humanin isoforms. *Genomics* 2009, 94, 247-256.

Bonawitz, ND; Chatenay-Lapointe, M; Wearn, CM; Shadel, GS. Expression of the rDNA-encoded mitochondrial protein Tar1p is stringently controlled and responds differentially to mitochondrial respiratory demand and dysfunction. *Curr Genetics* 2008, 54, 83-94.

Chrzanowska-Lightowlers, ZMA; Pajak, A; Lightowlers, RN. Termination of protein synthesis in mammalian mitochondria. *J Biol Chem* 2011, 286, 34479-34485.

Clayton, DA. Transcription of the mammalian mitochondrial genome. *Annu Rev Biochem* 1984, 53, 573-594.

Coelho, PSR; Bryan, AC; Kumar, A; Shadel, GS; Snyder, M. A novel mitochondrial protein, Tar1p, is encoded on the antisense strand of the nuclear 25S rDNA. *Genes Develop* 2002, 16, 2755-2760.

Eberhardt, RY; Haft, DH; Punta, M; Martin, M; O'Donovan, C; Bateman A. AntiFam: a tool to help identify spurious ORFs in protein annotation. *Database* 2012, bas003.

Faure, E; Delaye, L; Tribolo, S; Levasseur, A; Seligmann, H; Barthélémy, R-M. Probable presence of an ubiquitous cryptic mitochondrial gene on the antisense strand of the cytochrome oxidase I gene. *Biol Direct* 2011, 6, 56.

Galopier, A; Hermann-Le Denmat, S. Mitochondria of the yeasts *Saccharomyces cerevisiae* and *Kluyveromyces lactis* contain nuclear rDNA-encoded proteins. *PLoSone* 2011, 6, e16325.

Guo, B; Zhai, D; Cabezas, E; Welsh, K; Nouraini, S; Satterthwait, AC; Reed, JC. Humanin peptide suppresses apoptosis by interfering with Bax activation. *Nature* 2003, 423, 456-461.

Hashimoto, Y; Niikura, T; Tajima, H; Yasukawa, T; Sudo, H; Ito, Y; Kita, Y; Kawasumi, M; Kouyama, K; Doyu, M; Sobue, G; Koide, T; Tsuji, S; Lang, J; Kurokawa, K; Nishimoto, I. A rescue factor abolishing neuronal cell death by a wide spectrum of familial Alzheimer's disease genes and Abeta. *Proc Natl Acad Sci* U S A 2001, 98, 6336-6341.

Hausner, G; Monteiro-Vitorello, CB; Searles, DB; Maland, M; Fulbright, DW; Bertrand, H. A long open reading frame in the mitochondrial LSU rRNA group-I intron of *Cryphonectria parasitica* encodes a putative S5 ribosomal protein fused to a maturase. *Curr Genet* 1999, 35, 109-117.

Ingman, M; Gyllensten, U. mtDB: Human Mitochondrial Genome Database, a resource for population genetics and medical sciences. *Nuc Acids Res* 2006, 34, D749-D751.

Johansen, SD; Haugen, P; Nielsen, H. Expression of protein-coding genes embedded in ribosomal DNA. *Biol Chem* 2007, 388, 679-686.

Kermekchiev, M; Ivanova, L. Ribin, a protein encoded by a message complementary to rRNA, modulates ribosomal transcription and cell proliferation. *Mol Cell Biol.* 2001,21, 8255-8263.

Korostelev, AA. Structural aspects of translation termination on the ribosome. *RNA* 2011, 17, 1409-1421.

Krishnan, NM; Seligmann, H; Rao, BJ. Relationship between mRNA secondary structure and sequence variability in Chloroplast genes: possible life history implication. *BMC Genomics* 2008, 9, 48.

Lowe, TM; Eddy, SR. tRNAscan-SE: a program for improved detection of transfer RNA genes in genomic sequence. *Nuc Acids Res* 1997, 25, 955-964.

Pacheu-Grau, D; Gómez-Durán, A; López-Gallardo, E; Montoya, J; Ruiz-Pesini, E. *Read-through therapy for mitochondrial DNA nonsense mutations.* Drug Discover Today, in press.

Peng, G; Taylor, JD; Tchen, TT. Increased mitochondrial activities in pigmented (melanized) fish cells and nucleotide sequence of mitochondrial large rRNA. *Biochem Biophys Res Commun.* 1992, 189, 445-449.

Richter, R; Pajak, A; Dennerlein, S; Rozanska, A; Lightowlers, RN; Chrzanowska-Lightowlers, ZM. Translation termination in human mitochondrial ribosomes. *Biochem Soc Trans* 2011, 38, 1523-1526.

Ruiz-Pesini, E; Lott, MT; Procaccio, V; Poole, J; Brandon, MC; Mishmar, D; Yi, C; Kreuziger, J; Baldi, P; Wallace, DC. An enhanced MITOMAP with a global mtDNA mutational phylogeny. *Nuc Acids Res* 2007, 35 (Database issue), D823-D828. URL: http://www.mitomap.org.

Sabath, N; Graur, D. Detection of functional overlapping genes: simulation and case studies. *J Mol Evol* 2010, 7, 308-316.

Sabath, N; Landan, G; Graur, D. A method for the simultaneous estimation of selection intensities in overlapping genes. *PLoS ONE* 2008, 3, e3996.

Sabath, N; Price, N; Graur, D. A potentially novel overlapping gene in the genomes of Israeli acute paralysis virus and its relatives. *Virol* J 2009, 6, 144.

Schattner, P; Brooks, AN; Lowe, TM. The tRNAscan-SE, snoscan and snoGPS web servers for the detection of tRNAs and snoRNAs. *Nuc Acids Res 200*,5 33, W686-W689.

Schneider, A. Mitochondrial tRNA import and its consequences for mitochondrial translation. *Ann Rev Biochem* 2011, 80, 1033-1053.

Seligmann, H. Avoidance of antisense, antiterminator anticodons in vertebrate mitochondria. *Biosystems* 2010a, 101, 42-50.

Seligmann, H. Undetected antisense tRNAs in mitochondrial genomes*? Biol Direct* 2010b, 5, 39.

Seligmann, H. Do anticodons of misacylated tRNAs preferentially mismatch codons coding for the misloaded amino acid? *BMC Mol Biol* 2010c, 11, 41.

Seligmann, H. Two genetic codes, one genome: Frameshifted primate mitochondrial genes code for additional proteins in presence of antisense antitermination tRNAs. *Biosystems* 2011a ,105, 271-285.

Seligmann, H. Frameshifted chelonian mitochondrial genes code for additional proteins in presence of antisense antitermination tRNAs and the special case of *Lepidochelys*. Societas Europaea Herpetologica European Congress of Herpetology and Deutsche Gesellschaft fuer Herpetologie und Terrarienkunde Deutscher Herpetologentag, *Luxemburg und Trier,* 25-29 September 2011b, talk, abstract pages 85-86.

Seligmann, H. Pathogenic mutations in antisense mitochondrial tRNAs. *J Theor Biol* 2011c, 269, 287-296.

Seligmann, H. Error compensation of tRNA misacylation by codon-anticodon mismatch prevents translational amino acid insertion. *Comp Biol Chem* 2011d, 35, 81-95.

Seligmann, H. An overlapping genetic code for frameshifted overlapping genes in *Drosophila* mitochondria: Antisense antitermination tRNAs UAR insert serine. *J Theor Biol* 2012a, 298, 51-76.

Seligmann, H. Coding constraints modulate chemically spontaneous mutational replication gradients in mitochondrial genomes. *Current Genomics* 2012b, 13, 37-54.

Tabak, HF; Grivell, LA; Borst, P; Attardi, G. Transcription of mitochondrial DNA. *Crit Rev Biochem Mol Biol* 1983, 14, 297-317.

Temperley, R; Richter, R; Dennerlein, S; Lightowlers, RN; Chrzanowska-Lightowlers, ZM. Hungry codons promote frameshifting in human mitochondrial ribosomes. *Science* 2010, 327, 301–1301.

Tripp, HJ; Hewson, I; Boyarsky, S; Stuart, JM; Zehr, JP. Misannotations of rRNA can now generate 90% false positive protein matches in metatranscriptomic studies. *Nuc Acids Res* 2011, 39, 8792-8802.

Wei, Y; Thompson, J; Floudas, CA. CONCORD: a consensus method for protein secondary structure prediction via Mixed Integer Linear Optimization. *Proc Roy Soc* A 2012, 468, 831-850.

Young, DJ; Edgar, CD; Murphy, J; Fredebohm, J; Poole, ES; Tate, WP. Bioinformatic, structural, and functional analyses support release factor-like MTRF1 as a protein able to decode nonstandard stop codons beginning with adenine in vertebrate mitochondria. *RNA* 2010, 16, 1146-1155.

In: Ribosomes
Editors: Zhou Lin and Wang Liu

ISBN: 978-1-62417-698-2

Chapter 5

Translational Apparatus of Microorganisms at Low Temperature

***Jihwan Hwang*[1], *Masayori Inouye*[2] *and Sangita Phadtare*[2*]**
[1]Department of Microbiology, Pusan National University, Busan, Republic of Korea
[2]Department of Biochemistry and Molecular Biology, Robert Wood Johnson Medical School, UMDNJ, CABM, Piscataway, NJ, US

Abstract

Temperature up-shift or down-shift is one of the major environmental stresses encountered by microorganisms. Especially, one of the main consequences of temperature down-shift on cells is conformational changes of secondary structures of RNAs. These cold-shock induced structural changes of RNAs have significant effect on the global gene expressions, i) by improper termination of transcription, ii) inefficient RNA degradation, iii) translation of alternatively-structured mRNAs, and iv) biogenesis of ribosome and processing of rRNA. Several cold-shock proteins are produced to counteract these effects and thus allow cold acclimatization of the cell. This chapter describes how specific cold shock proteins modulate the ribosome biogenesis and eventually translation, to ensure effective protein production at low temperature.

Introduction

Living organisms encounter several stresses such as extreme pH, pressure *etc.* Temperature, either significantly higher (heat shock) or lower (cold shock) than their optimal growth temperature is one such common stress. In order to understand how microorganisms cope with low temperature stress, it is important to study the cold-shock response and

* Corresponding author. Mailing address: Department of Biochemistry and Molecular Biology, Robert Wood Johnson Medical School, University of Medicine and Dentistry of New Jersey, Center for Advanced Biotechnology and Medicine, 679 Hoes Lane, Piscataway, New Jersey, 08854, US. Phone: (732) 235-3912, Fax: (732) 235-4559, E-mail: phadtasa@umdnj.edu.

adaptation. *Escherichia coli* and *Bacillus subtilis* have been used as model systems to uncover major principles of cold-shock response and adaptation (for a recent review see Phadtare & Severinov 2010). Enterobacteria such as *E. coli* constitute the normal gut flora of humans. These are often excreted in faces and as a result, encounter sudden drastic temperature downshift. As this bacterium is equipped with efficient cold-shock response machinery, it is able to quickly adapt to its new environment after excretion. This cellular machinery consists of cold shock proteins, which are produced after the temperature downshift and help the cells counteract the detrimental effects of cold shock. In this chapter, we will discuss how cold shock affects different aspects of cell metabolism and machinery with focus on translational apparatus of *E. coli* and how cold shock proteins counteract the detrimental effects of temperature downshift to aid cellular adaptation to cold.

The effects of temperature downshift on the cell are varied and mainly affect cellular transport, transcription and translation. Some of these major cellular changes include (i) decrease in membrane fluidity, (ii) changes in nucleic acid structures, (iii) inefficient protein folding and (iv) reduced ribosome function which affects translation, especially of non-cold-shock proteins (for a review see Klinkert & Narberhaus 2009).

Cell membranes have a liquid crystalline nature under optimal temperature conditions. This state changes to a gel-phase state in response to temperature downshift. This change adversely affects active transport and protein secretion. The cells counteract this detrimental change by increasing the proportion of unsaturated fatty acids (UFAs) in the membrane lipids. Phospholipids with UFAs have lower melting points and greater degree of flexibility than phospholipids containing saturated fatty acids (Sinensky 1974). In *E. coli,* the UFA produced after temperature downshift is *cis*-vaccenic acid. The enzyme β-ketoacyl-acyl carrier protein (ACP) synthase II is activated by low temperature and converts palmitoleic acid to *cis*-vaccenic acid (Garwin & Cronan 1980; Garwin *et al.* 1980). Bacteria such as *Bacillus subtilis* and photosynthetic cyanobacteria contain membrane-bound desaturases encoded by *des*. Both the synthesis and stability of desaturase increase in response to temperature downshift (Murata & Wada 1995; Aguilar *et al.* 1998). It has been shown that this enzyme inserts a double bond into the acyl chain of membrane phospholipids. Food-borne pathogen such as *Listeria monocytogenes* adapts another mechanism for compensating for the temperature-induced change in membrane composition. Upon temperature downshift, the membranes for this bacterium predominantly contain branched-chain fatty acids, especially the *anteiso* fatty acids. These fatty acids have lower melting point than *iso* fatty acids and thus play a critical role in maintaining a fluid, liquid-crystalline state of the membrane lipids (Annous *et al.* 1997).

Effect of cold shock on nucleic acids is two-fold, (a) negative supercoiling of DNA increases and (b) secondary structures in RNA and DNA are stabilized. The increase in negative supercoiling of DNA affects replication, transcription, and recombination (Wang & Syvanen 1992; Krispin & Allmansberger 1995; Mizushima *et al.* 1997). DNA is usually negatively supercoiled and the extent of the supercoil is regulated mainly by DNA gyrase and DNA topoisomerase I. The changes in the number of base pairs per helical turn are due to a change in superhelical density. As recognition of some σ^{70} promoters by RNA polymerase is dependent on relative orientation of the –35 and –10 regions, the changes in supercoiling affect transcription (Wang & Syvanen 1992). RNA structures are sensitive to temperature changes and thus have been proposed to act as cellular thermometers (Storz 1999; Narberhaus *et al.* 2006; Klinkert & Narberhaus 2009). Stabilization of the secondary structures in RNAs due to low temperature affects transcription and translation due to hindered movement of RNA polymerase and ribosomes, respectively. Certain *E. coli* proteins such as CspA and its homologues act as RNA chaperones (Jiang *et al.* 1997; Bae *et al.* 2000) by melting the secondary structures in nucleic acids and thus aiding transcription and translation at low temperature. This activity of CspA and its homologues is essential for cold acclimation of cells (Phadtare *et al.* 2002a; Phadtare & Inouye 2004a).

Although not severe as that seen with heat-shock, downshift of temperature also leads to protein misfolding (Kandror & Goldberg 1997). Global transcript analysis of cold-shocked *E. coli* cells showed that certain molecular chaperones such as Caseinolytic proteases (Clps), and trigger factor (TF) (Phadtare & Inouye 2004a) were induced upon cold shock. Chaperones which are critical at heat shock such as GroEL and GroES are also induced to some extent after cold shock. (Kandror *et al.* 2002). ClpB renatures and solubilizes aggregated proteins at low temperatures (Porankiewicz & Clarke 1997). TF induced by cold shock at a modest level (Kandror *et al.* 2002) is a peptidyl prolyl isomerase that catalyzes the *cis/trans* isomerization of peptide bonds which are N-terminal to the proline residue (Kandror & Goldberg 1997). TF plays an important role in co-translational protein folding (Blaha *et al.* 2003; Maier *et al.* 2003). Temperature downshift also leads to changes such as reversible methylation of cytoplasmic signaling/adaptation domain of aspartate chemoreceptor (Tar) of *E. coli* (Nishiyama *et al.* 1999).

Ribosomes act as the sensors of temperature in bacteria as the state of ribosomes signals the induction of the temperature responses (VanBogelen & Neidhardt 1990). It was proposed (Graumann & Marahiel 1996) that after an increase in temperature, the translation takes place at a rate higher than that at which charged tRNA are supplied, hence the A-site of the ribosome is empty. After a temperature downshift however, the translational efficiency is decreased, so the A-site is blocked due to the high concentration of charged tRNA. This in turn lowers (p)ppGpp [guanosine 5'triphosphate-3'diphosphate (pppGpp) and guanosine 5'diphosphate-3'diphosphate (ppGpp) (collectively abbreviated as (p)ppGpp)] levels by inhibiting RelA mediated synthesis of (p)ppGpp. Thus, the concentration of (p)ppGpp increases at high temperatures and decreases after a temperature downshift. It has been shown that artificially induced high levels of (p)ppGpp diminish the expression of cold-shock proteins, while low concentration increases their production (Jones *et al.* 1992; Jones & Inouye 1994). Thus, (p)ppGpp affects the magnitude of the cold-shock response. It has been suggested that upon cold shock, ribosomes become nonfunctional for cellular mRNAs, except for mRNAs for cold-shock proteins (Jones & Inouye 1996). The mRNAs for genes encoding cold shock proteins contain certain structural elements such as translation enhancing element (Mitta *et al.* 1997; Moll *et al.* 2001; Qing *et al.* 2004), which helps to reduce the detrimental effect of cold shock on ribosomes by facilitating translation initiation. This leads to the production of cold-shock ribosome factors such as RbfA and CsdA during the lag period of growth after the temperature downshift. These proteins then bind ribosomes and convert them to cold-adapted ribosomes that are capable of translating non-cold-shock mRNAs (Jones & Inouye 1996).

Cold shock also leads to induction of several genes encoding proteins involved in sugar transport and metabolism (Phadtare & Inouye 2004a). These sugars include maltose, mannose, ribose, xylose, and trehalose. Sugars are shown to confer protection to the cells undergoing cold shock. The cold-shock induction of mannose and maltose transport systems was prominently repressed in a cold-sensitive *csp* quadruple deletion mutant (Xia *et al.*

2001b) that has a significantly prolonged (4 h) lag period as opposed to 1 h lag period of the wild-type strain at 15°C. Trehalose is previously known to confer protection to cells against heat, cold and osmotic stress by deterring denaturation and aggregation of proteins, protecting against oxidative damage, and also by stabilizing cellular membrane (Kandror *et al.* 2002).

Our focus in this chapter is to discuss how specific cold shock proteins modulate the ribosome biogenesis and eventually translation, to ensure effective protein production at low temperature.

Proteins Associated with Ribosomal Function

Proteins that significantly modulate the ribosome biogenesis and function at low temperature include RNA helicase, CsdA, RbfA and cold shock protein, CspA (Figure 1).

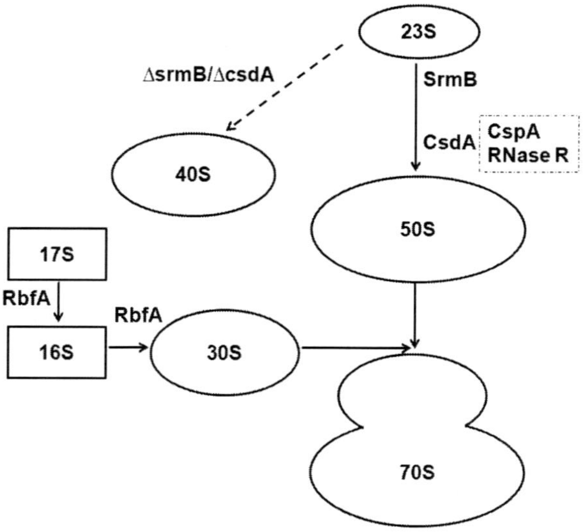

Figure 1. Schematic representation of the role of cold shock proteins involved in ribosome biogenesis at low temperature.

(i) CsdA

CsdA is a multifunctional, DEAD-box protein which is essential only at low temperature (Jones *et al.* 1996; Charollais *et al.* 2004). *csdA* deletion cells exhibit cold-sensitive phenotype at low temperatures characterized by decreased polysomes (Jones *et al.* 1996) and accumulation of 40S particles (Peil *et al.* 2008; Jagessar & Jain 2010). CsdA belongs to the family of putative RNA helicases conserved from bacteria to humans, which are known to play important roles in many cellular processes such as processing, transport or degradation of RNA or ribosome biogenesis (Iost & Dreyfus 2006; Rajkowitsch *et al.* 2007). There are five DEAD-box genes, *csdA*, *dbpA*, *rhlB*, *rhlE* and *srmB* in *E. coli* (Linder *et al.* 1989). DEAD-box proteins are characterized by presence of at least nine conserved amino acid motifs in their core, one of which is D-E-A-D box (Linder *et al.* 1989; Tanner *et al.* 2003).

Based on the data accumulated from different groups, multiple functions have been attributed to CsdA such as, (i) Ribosome biogenesis- the *deaD/csdA* gene was originally identified as a multicopy suppressor of a mutation in the *rpsB* gene that encodes r-protein. This suggested that CsdA may play a role in the biogenesis of the small ribosomal subunit (Toone *et al.* 1991). Later on it was reported that over-expression of CsdA in an S2 mutant restores the incorporation of r-proteins S2 and S1 into the 30S ribosomal subunits (Moll *et al.* 2002). Recently, it was shown that CsdA is involved in the biogenesis of the 50S ribosomal subunits and deletion of the *csdA* gene leads to a deficit in free 50S subunits and accumulation of a 40S-like particle. Most of the ribosomal proteins that are present in reduced amounts in the precursor particles are assembled at a late stage in the assembly process, suggesting that CsdA is involved at a stage in 50S biogenesis later than that of SrmB (Charollais *et al.* 2004; Shajani *et al.* 2011), although it is possible that CsdA may also participate in more than one assembly steps (Peil *et al.* 2008). The authors also showed that CsdA associates with 50S precursors at low temperature and can complement the ribosome defect of *srmB* deletion strain (Charollais *et al.* 2004). (ii) Translation initiation- CsdA has homology to the eukaryotic initiation factor eIF4A that catalyzes the ATP-dependent unwinding of RNA duplexes and stimulates translational initiation in eukaryotic cells. Thus, it was suggested that CsdA may assist in translation by promoting translation initiation of structured mRNAs (Lu *et al.* 1999). (iii) Degradation of mRNAs- It has been shown that CsdA accumulates during the early stages of cold acclimatization and assembles into degradosomes with RNase E (Yamanaka & Inouye 2001; Khemici *et al.* 2004; Prud'homme-Genereux *et al.* 2004). It was also shown that for continued growth of the cold-acclimated cells at low temperature, it is beneficial if the massive production of CspA and its homologues immediately following the temperature downshift be transient. Both PNPase and CsdA are important for this to happen. CsdA, by virtue of its helicase activity can unwind the secondary structures in mRNAs to some extent. Indeed it was shown to be involved in the efficient and selective degradation of Csp mRNAs by unwinding the mRNA secondary structure that impedes the processive activity of PNPase (Yamanaka & Inouye 2001).

To further elucidate cold-shock function of CsdA, we screened for multicopy suppressor(s) for the cold-sensitive phenotype of the *csdA* deletion cells at 15°C. We observed that RhlE can complement CsdA function at low temperature. This finding suggested that the helicase activity of CsdA is important for its function that is critical for cold acclimation of cells. This was further supported by the observation that the DEAD-box amino acids that are critical for the helicase activity of CsdA are also essential for its cold-acclimation activity (Awano *et al.* 2007; Turner *et al.* 2007). We also observed that, two cold-shock-inducible proteins, an RNA chaperone, CspA, and an exonuclease, RNase R can also complement cold-shock function of CsdA. We also observed that absence of CsdA and RNase R results in increased sensitivity of the cells to moderate temperature downshifts. We tested the correlation between the helicase activity of CsdA and stability of mRNAs of cold-inducible genes using *cspA* mRNA, which was significantly stabilized in the Δ*csdA* cells. This stabilization was reversed by the overexpression of wild-type CsdA or RNase R, but not by the overexpression of helicase-deficient mutant of CsdA. These data suggested that the involvement of CsdA in the low temperature mRNA decay may be its primary role and its helicase activity is crucial for promoting degradation of mRNAs stabilized at low temperature (Awano *et al.* 2007). This helped us to put forth a hypothesis that RNA metabolism may be different at low temperature as compared to that at 37°C as the secondary structures in mRNAs are stabilized at low temperature and may need assistance of proteins such as CsdA to unwind these structures before they can be digested by ribonucleases (Phadtare 2011). Previously, it was reported that CsdA does not have strong specificity for RNA substrates *in vitro*; however, it may have substrate preferences *in vivo* (Iost & Dreyfus 2006). We thus carried out analysis of the global transcript pattern of cells deprived of or overexpressing CsdA at low temperature to elucidate how RNA metabolism is mediated at low temperature. The analysis revealed distinct sets of genes, which are possible targets of CsdA (Phadtare 2012).

(ii) SrmB

The *srmB* gene encoding SrmB, another DEAD-box RNA helicase, was originally isolated as multicopy suppressor of mutations in ribosomal protein genes *rplX* (encoding L24) (Nishi *et al.* 1988), suggesting that SrmB is involved in ribosome biogenesis. Deletion of the *srmB* gene leads to fewer 70S particles, decreased amount of free 50S ribosomal subunits and the accumulation of 40S particles. This leads to slower growth at low temperature. Analysis of the ribosomal RNA and protein contents of the 40S particle indicates that it represents an incompletely assembled large subunit. These particles showed decrease in amounts of ribosomal proteins L6, L7/L12, L14, L16, L25, L27, L31, L32, and L33, and were missing ribosomal proteins L13, L28, and L34 (Charollais *et al.* 2003). As L13 is essential for the formation of the first 50S *in vitro* intermediate (55), SrmB is thought to have a role in the early steps of ribosome biogenesis such as an early step of 50S assembly that is necessary for the binding of L13 (Charollais *et al.* 2003; Trubetskoy *et al.* 2009; Shajani *et al.* 2011). This step may consist of a structural rearrangement that, at low temperature, cannot occur without the assistance of SrmB (Charollais *et al.* 2003). This notion was further supported by the observation that SrmB forms a quaternary complex with the ribosomal proteins L4 and L24, which are also essential for the first 50S *in vitro* intermediate and a fragment from the 5' end of 23S rRNA that contains the binding site for L4 and L24 (Trubetskoy *et al.* 2009). Thus, SrmB clamps the 23S rRNA in a stable conformation for this assembly to be completed. Recently, a genetic screen was used to determine the site(s) of action of SrmB. Mutations that compensate for the absence of SrmB with respect to both the growth phenotype and the ribosome assembly defect, were located in a heptanucleotide at two nearby locations of the 23S rRNA domain II and in a complementary hexanucleotide in the 5S rRNA. These mutations disrupt the complementarity between 23S and 5S rRNA molecules and affect the ability to form intermolecular base-pairing. This suggested that SrmB may prevent misfolding of both 23S and 5S rRNA around the regions in which mutations were found. Complementary regions of 23S rRNA 1022–1027 and 5S rRNA 25–30 are not involved in Watson-Crick base-pairing but are parts of internal loops in the mature ribosome (Proux *et al.* 2011; Remme 2011). This suggests that SrmB may be resolving improper folding of the 23S and 5S rRNAs that would result in the inhibition of 50S subunit assembly. These misfolded conformations may get stabilized at low temperature, thus requiring activity of SrmB.

As the temperature decreases, the thermal energy available for rRNA fluctuations also decreases, which increases the possibility of rRNA being more likely to be trapped in misfolded states. CsdA and SrmB supposedly rescue the ribosomes from these misfolded states, although the exact molecular basis of this phenomenon is not resolved. The overexpression of SrmB does not suppress the cold sensitivity of the *csdA* deletion mutant (Charollais *et al.* 2004) and the phenotypes for the *srmB* and *deaD* deletions are not additive. This suggests that these two DEAD-box proteins may be involved in different stages of 50S biogenesis. Both proteins do however seem to share some common properties. Both proteins (i) can unwind duplexes with 3' or 5' extensions, (ii) overexpression of SrmB stabilizes certain mRNAs (Iost & Dreyfus 1994) and (iii) SrmB can also bind to RNase E (Khemici *et al.* 2004). It has been suggested that by virtue of their helicase activity, these proteins may assist 50S assembly by modulating RNA structures by facilitating structural transitions within the RNA and/or to allow proper binding of r-protein(s). These proteins may act as RNA chaperones that prevent and/or resolve misfolding. For example, rRNA may become trapped in incorrect structures, and require assistance to reach its active conformation (Herschlag 1995).

(iii) RbfA

Ribosome binding factor RbfA is one of the main cold shock proteins in *E. coli* (Dammel & Noller 1995). Recently, the three-dimensional structure of RbfA was solved by NMR spectroscopy (Kobayashi *et al.* 2012). Structure of RbfA from *Thermotoga maritima* has also been solved (Grimm & Wohnert 2005). *E. coli* RbfA contains a $\alpha+\beta$ fold containing three helices and three β strands. The structure has type-II KH-domain fold topology and is most similar to the KH domain of the *E. coli* Era GTPase. Its electrostatic field distribution is most similar to the KH1 domain of the NusA protein from *T. maritima*, another cold-shock associated RNA-binding protein.

RbfA was originally identified as a multicopy suppressor of a point mutation in the 5′-end helix of the 16S ribosomal RNA and was also shown to be associated with the 30S ribosomal subunit (Dammel & Noller 1995). A later study showed it to be required for efficient 16S rRNA processing (Bylund *et al.* 1998). RbfA is required for efficient processing of 17S rRNA to 16S rRNA and may also be involved in maturation after the formation of 16S rRNA by virtue of its binding to the 5' terminal helix region of mature 16S rRNA (Lovgren *et al.* 2004). Deletion of RbfA leads to loss of ability of the cell to adapt to low temperature and this leads to constitutive cold-shock response (Jones & Inouye 1996). The cold sensitivity of *rbfA* deletion cells is directly related to their lack of translation initiation-capable 30S subunits containing mature 16S rRNA at low temperature (Xia *et al.* 2003).

(iv) CspA and its Homologues

Several cold-shock inducible genes were identified by global transcript analysis of *E. coli* (Gualerzi *et al.* 2003; Phadtare & Inouye 2004b). CspA is the major cold-shock protein of *E. coli* and its homologues are found to be wide-spread among psychrophilic, psychrotrophic, mesophilic and thermophilic bacteria. *E. coli* has a family of nine CspA homologues (CspA to CspI), out of which four (CspA, CspB, CspG and CspI) are induced by cold shock. These exhibit functional redundancy (Xia *et al.* 2001b) and play a pivotal role in the cold-shock response and adaptation of bacteria. CspA homologues act as RNA chaperones (Jiang *et al.* 1997; Bae *et al.* 2000; Phadtare *et al.* 2002a). The secondary structures in RNA are stabilized at low temperatures and affect both transcription elongation as well as ribosomal movement on RNA and, thus, translation. CspA homologues bind and melt secondary structures in nucleic acids, thereby facilitating transcription and translation at low temperature. The RNA chaperone activity enables these proteins to act as transcription antiterminators at *rho*-independent terminators (Bae *et al.* 2000; Phadtare *et al.* 2002a; Phadtare *et al.* 2002b; Phadtare *et al.* 2004; Phadtare & Severinov 2005). This activity is critical for the cold acclimation of cells (Phadtare *et al.* 2002a) and is conserved in CspA homologues from higher organisms; for example, wheat Csp protein WCSP1 (Nakaminami *et al.* 2006).

The cold-shock induction of CspA homologues is presumed to be regulated at the levels of mRNA stability, translation and transcription and does not need additional transcription factors. Out of these, regulation at the level of transcription is not considered to be significant. A robust and transient stabilization of the *cspA* mRNA immediately following temperature downshift is mainly responsible for its cold-shock induction. The structural element that is responsible for this phenomenon is the untranslated 5' region of *cspA* mRNA. Another structural element in *cspA* mRNA plays an important role in its cold-shock translation, namely the translation enhancing element located 14 bases downstream of the initiation codon (Mitta *et al.* 1997; Moll *et al.* 2001; Qing *et al.* 2004). This element is also present in other cold-shock-inducible genes such as *cspB*, *cspG*, *cspI*, *csdA* and *rbfA*. This sequence is complementary to a region in the penultimate stem of 16S rRNA. The exact mechanism of the enhancing effect on translation initiation by this element is unknown at present. Recent evidences imply that the transcripts of *cspA* at low temperature act as a thermometer or thermoswitch for its restricted expression at low temperature by rearranging the overall secondary structure within the open reading frame (ORF) of transcript rather than within a long 5'-*UTR* (untranslated region) (Giuliodori *et al.* 2010). It is likely that the ordered structure of *cspA* ORF is resistant to the RNase activity and that free 5'-*UTR* is involved in the translation regulation of *cspA* mRNA as described below in "LACE" (the low temperature-dependent antibiotic effect of truncated *cspA* expression) phenotype (Jiang *et al.* 1996b; Jiang *et al.* 1996c; Xia *et al.* 2001a).

The LACE Effect

We observed that expression of mutant *cspA* mRNA encoding a premature nonsense codon from a high copy number vector at low temperature led to the accumulation of most of the cellular ribosomes on mutant *cspA* transcripts, ultimately causing cell death (Jiang *et al.* 1996a). This phenomenon is termed the low temperature-dependent antibiotic effect of truncated *cspA* expression (LACE). The LACE caused by nonsense mutations in the *cspA* gene results from a nearly exclusive trapping of ribosomes by the mutant *cspA* mRNAs overexpressed at low temperature. Such trapping occurs even with a nonsense mutation at the second codon, indicating that the LACE can occur in the absence of translation elongation. This in turn suggests that translation initiation of the *cspA* mRNA is highly efficient and is responsible for its outstanding translatability. Interestingly, depending on the position of non-sense codon, those accumulated ribosomes were stalled at the premature stop codon yielding monosome, disomes, trisomes or tetrasomes. Not only the C-terminally truncated mutant but also the substitutions of aromatic amino acid residues on the surface of CspA protein on hydrophobic patch caused the LACE effect. However, disruptions of two *cis*-elements in the 5'-*UTR*, 'Upstream Box' and Shine-Dalgarno (SD) sequence that are complementary to the 16S rRNA sequences did not show growth inhibition (Yamanaka *et al.* 1999; Xia *et al.*

2001a). Therefore, it was suggested that in the absence of functional CspA, the overexpressed highly translatable *cspA* transcripts result in the accumulation of translating ribosomes selectively at premature stop codon. (Xia *et al.* 2001a). Based on these observations it was proposed that CspA proteins expressed upon cold-shock may modulate the duplex formation between regulatory *cis*-elements and 16S rRNA sequences probably reducing the production of CspA post acclimation phase. Interestingly, an approach to elucidate the mechanism of the LACE effect using screening of genomic library revealed that functionally overlapping ATP-dependent DNA helicases, UvrD and DinG can suppress the LACE phenotype recovering the cell growth at low temperature (Hwang *et al.* 2012). This suppression effect was dependent on nucleic acid- and ATP-binding activities of these helicases. Among fourteen DNA helicases, UvrD and DinG represent two different DNA helicase superfamily 1 and 2, respectively, and both translocate in opposite directions on nucleic acids (Tuteja & Tuteja 2004). DNA-damage recruits UvrD to carry out nucleotide excision repair and mismatch repair (Runyon et al. 1990; Tuteja & Tuteja 2004). Rather, DinG is an SOS-regulated helicase translocating in the 5'→3' direction on ssDNA (Lewis et al. 1992; Courcelle et al. 2001; Tuteja & Tuteja 2004). Previously, both UvrD and DinG were shown to unwind DNA·DNA duplexes as well as DNA·RNA hybrids *in vitro* (Matson 1989; Voloshin & Camerini-Otero 2007), and more recently, it was reported that either UvrD or DinG in a conjunction with Rep cooperatively promotes DNA replication where actively transcribing RNA polymerase head-on collides, resulting in the inhibition of transcription (Boubakri et al. 2010). Even though *cspA* is not considered to be hyperactive transcription unit, it is possible that UvrD or DinG negatively regulate the transcriptional activity of the *cspA* gene, whose transcriptional direction on *E. coli* genome collides into DNA replication. In addition, one may speculate that the nascent *cspA* transcripts are dissociated from RNA polymerase, or that duplex formation between 16S rRNA fragment sequence and *cis*-element of *cspA* 5'-UTR was interrupted by those DNA helicases, eventually leading to reduction of *cspA* expression. Lastly, the N-terminal domain of *Bacillus subtilis* DinG shows a homology to an exonuclease domain (Moser *et al.* 1997), thus, it is also possible that DinG may modulate the metabolism of *cspA* mRNAs in LACE suppression system. As described earlier, RbfA, RNA helicases such as CsdA and SrmB or CspA have RNA binding domain and perform their unique role in either mRNA translation, rRNA processing or ribosome biogenesis at low temperature. However, identification of two DNA helicases as suppressors for LACE is somewhat surprising; nevertheless, it may present a novel possibility that both DNA helicases are involved either directly or indirectly in the gene expression under cold-shock stress condition.

Conclusion

Cold shock affects several cellular functions, translation being one of these. Stabilization of secondary structures in mRNAs is one of the main factors responsible for hindrance of translation at low temperature. A number of cold shock proteins such as RNA chaperones and helicases with unwinding activities are thus important for facilitation of translation leading to cold acclimation of cells.

References

Aguilar, P.S., Cronan, J.E., Jr. & de Mendoza, D. (1998) A *Bacillus subtilis* gene induced by cold shock encodes a membrane phospholipid desaturase. *Journal of bacteriology* 180, 2194-2200.

Annous, B.A., Becker, L.A., Bayles, D.O., Labeda, D.P. & Wilkinson, B.J. (1997) Critical role of anteiso-C15:0 fatty acid in the growth of *Listeria monocytogenes* at low temperatures. *Applied and environmental microbiology* 63, 3887-3894.

Awano, N., Xu, C., Ke, H., Inoue, K., Inouye, M. & Phadtare, S. (2007) Complementation analysis of the cold-sensitive phenotype of the *Escherichia coli csdA* deletion strain. *Journal of bacteriology* 189, 5808-5815.

Bae, W., Xia, B., Inouye, M. & Severinov, K. (2000) *Escherichia coli* CspA-family RNA chaperones are transcription antiterminators. *Proceedings of the National Academy of Sciences of the United States of America* 97, 7784-7789.

Blaha, G., Wilson, D.N., Stoller, G., Fischer, G., Willumeit, R. & Nierhaus, K.H. (2003) Localization of the trigger factor binding site on the ribosomal 50S subunit. *Journal of molecular biology* 326, 887-897.

Boubakri, H., de Septenville, A.L., Viguera, E. & Michel, B. (2010) The helicases DinG, Rep and UvrD cooperate to promote replication across transcription units *in vivo*. *EMBO Journal* 29, 145-157.

Bylund, G.O., Wipemo, L.C., Lundberg, L.A. & Wikstrom, P.M. (1998) RimM and RbfA are essential for efficient processing of 16S rRNA in *Escherichia coli*. *Journal of bacteriology* 180, 73-82.

Charollais, J., Dreyfus, M. & Iost, I. (2004) CsdA, a cold-shock RNA helicase from *Escherichia coli*, is involved in the biogenesis of 50S ribosomal subunit. *Nucleic acids research* 32, 2751-2759.

Charollais, J., Pflieger, D., Vinh, J., Dreyfus, M. & Iost, I. (2003) The DEAD-box RNA helicase SrmB is involved in the assembly of 50S ribosomal subunits in *Escherichia coli*. *Molecular microbiology* 48, 1253-1265.

Courcelle, J., Khodursky, A., Peter, B., Brown, P.O. & Hanawalt, P.C. (2001) Comparative gene expression profiles following UV exposure in wild-type and SOS-deficient *Escherichia coli*. *Genetics* 158, 41-64.

Dammel, C.S. & Noller, H.F. (1995) Suppression of a cold-sensitive mutation in 16S rRNA by overexpression of a novel ribosome-binding factor, RbfA. *Genes & development* 9, 626-637.

Garwin, J.L. & Cronan, J.E., Jr. (1980) Thermal modulation of fatty acid synthesis in *Escherichia coli* does not involve *de novo* enzyme synthesis. *Journal of bacteriology* 141, 1457-1459.

Garwin, J.L., Klages, A.L. & Cronan, J.E., Jr. (1980) Beta-ketoacyl-acyl carrier protein synthase II of *Escherichia coli*. Evidence for function in the thermal regulation of fatty acid synthesis. *The Journal of biological chemistry* 255, 3263-3265.

Giuliodori, A.M., Di Pietro, F., Marzi, S., Masquida, B., Wagner, R., Romby, P., Gualerzi, C.O. & Pon, C.L. (2010) The *cspA* mRNA Is a Thermosensor that Modulates Translation of the Cold-Shock Protein CspA. *Molecular Cell* 37, 21-33.

Graumann, P. & Marahiel, M.A. (1996) Some like it cold: response of microorganisms to cold shock. *Archives of microbiology* 166, 293-300.

Grimm, S.K. & Wohnert, J. (2005) NMR assignments of the cold-shock protein ribosome-binding factor A (RbfA) from *Thermotoga maritima*. *Journal of Biomolecular NMR* 31, 73-74.

Gualerzi, C.O., Giuliodori, A.M. & Pon, C.L. (2003) Transcriptional and post-transcriptional control of cold-shock genes. *Journal of Molecular Biology* 331, 527-539.

Herschlag, D. (1995) RNA chaperones and the RNA folding problem. *The Journal of biological chemistry* 270, 20871-20874.

Hwang, J., Lee, K., Phadtare, S. & Inouye, M. (2012) Identification of Two DNA Helicases UvrD and DinG as Suppressors for Lethality Caused by Mutant *cspA* mRNAs. *Journal of molecular microbiology and biotechnology* 22, 135-146.

Iost, I. & Dreyfus, M. (1994) mRNAs can be stabilized by DEAD-box proteins. *Nature* 372, 193-196.

Iost, I. & Dreyfus, M. (2006) DEAD-box RNA helicases in *Escherichia coli*. *Nucleic acids research* 34, 4189-4197.

Jagessar, K.L. & Jain, C. (2010) Functional and molecular analysis of *Escherichia coli* strains lacking multiple DEAD-box helicases. *RNA (New York, N.Y)* 16, 1386-1392.

Jiang, W., Fang, L. & Inouye, M. (1996a) Complete growth inhibition of *Escherichia coli* by ribosome trapping with truncated *cspA* mRNA at low temperature. *Genes to Cells* 1, 965-976.

Jiang, W., Hou, Y. & Inouye, M. (1997) CspA, the major cold-shock protein of *Escherichia coli*, is an RNA chaperone. *The Journal of biological chemistry* 272, 196-202.

Jiang, W.N., Fang, L. & Inouye, M. (1996b) The role of the 5'-end untranslated region of the mRNA for CspA, the major cold-shock protein of *Escherichia coli*, in cold-shock adaptation. *Journal of bacteriology* 178, 4919-4925.

Jiang, W.N., Li, F. & Inouye, M. (1996c) Complete growth inhibition of *Escherichia coli* by ribosome trapping with truncated *cspA* mRNA at low temperature. *Genes to Cells* 1, 965-976.

Jones, P.G., Cashel, M., Glaser, G. & Neidhardt, F.C. (1992) Function of a relaxed-like state following temperature downshifts in *Escherichia coli*. *Journal of bacteriology* 174, 3903-3914.

Jones, P.G. & Inouye, M. (1994) The cold-shock response--a hot topic. *Molecular microbiology* 11, 811-818.

Jones, P.G. & Inouye, M. (1996) RbfA, a 30S ribosomal binding factor, is a cold-shock protein whose absence triggers the cold-shock response. *Molecular microbiology* 21, 1207-1218.

Jones, P.G., Mitta, M., Kim, Y., Jiang, W. & Inouye, M. (1996) Cold shock induces a major ribosomal-associated protein that unwinds double-stranded RNA in *Escherichia coli*. *Proceedings of the National Academy of Sciences of the United States of America* 93, 76-80.

Kandror, O., DeLeon, A. & Goldberg, A.L. (2002) Trehalose synthesis is induced upon exposure of *Escherichia coli* to cold and is essential for viability at low temperatures. *Proceedings of the National Academy of Sciences of the United States of America* 99, 9727-9732.

Kandror, O. & Goldberg, A.L. (1997) Trigger factor is induced upon cold shock and enhances viability of *Escherichia coli* at low temperatures. *Proceedings of the National Academy of Sciences of the United States of America* 94, 4978-4981.

Khemici, V., Toesca, I., Poljak, L., Vanzo, N.F. & Carpousis, A.J. (2004) The RNase E of *Escherichia coli* has at least two binding sites for DEAD-box RNA helicases: functional replacement of RhlB by RhlE. *Molecular microbiology* 54, 1422-1430.

Klinkert, B. & Narberhaus, F. (2009) Microbial thermosensors. *Cellular and molecular life sciences* 66, 2661-2676.

Kobayashi, H., Swapna, G.V., Wu, K.P., Afinogenova, Y., Conover, K., Mao, B., Montelione, G.T. & Inouye, M. (2012) Segmental isotope labeling of proteins for NMR structural study using a protein S tag for higher expression and solubility. *Journal of biomolecular NMR* 52, 303-313.

Krispin, O. & Allmansberger, R. (1995) Changes in DNA supertwist as a response of *Bacillus subtilis* towards different kinds of stress. *FEMS microbiology letters* 134, 129-135.

Lewis, L.K., Jenkins, M.E. & Mount, D.W. (1992) Isolation of DNA damage-inducible promoters in *Escherichia coli*: regulation of *polB* (*dinA*), *dinG*, and *dinH* by LexA repressor. *Journal of bacteriology* 174, 3377-3385.

Linder, P., Lasko, P.F., Ashburner, M., Leroy, P., Nielsen, P.J., Nishi, K., Schnier, J. & Slonimski, P.P. (1989) Birth of the D-E-A-D box. *Nature* 337, 121-122.

Lovgren, J.M., Bylund, G.O., Srivastava, M.K., Lundberg, L.A., Persson, O.P., Wingsle, G. & Wikstrom, P.M. (2004) The PRC-barrel domain of the ribosome maturation protein RimM mediates binding to ribosomal protein S19 in the 30S ribosomal subunits. *RNA (New York, N.Y)* 10, 1798-1812.

Lu, J., Aoki, H. & Ganoza, M.C. (1999) Molecular characterization of a prokaryotic translation factor homologous to the eukaryotic initiation factor eIF4A. *International Journal of Biochemistry and Cell Biology* 31, 215-229.

Maier, R., Eckert, B., Scholz, C., Lilie, H. & Schmid, F.X. (2003) Interaction of trigger factor with the ribosome. *Journal of molecular biology* 326, 585-592.

Matson, S.W. (1989) Escherichia coli DNA helicase II (*uvrD* gene product) catalyzes the unwinding of DNA.RNA hybrids *in vitro*. *Proceedings of the National Academy of Sciences of the United States of America* 86, 4430-4434.

Mitta, M., Fang, L. & Inouye, M. (1997) Deletion analysis of *cspA* of *Escherichia coli*: requirement of the AT-rich UP element for *cspA* transcription and the downstream box in the coding region for its cold shock induction. *Molecular microbiology* 26, 321-335.

Mizushima, T., Kataoka, K., Ogata, Y., Inoue, R. & Sekimizu, K. (1997) Increase in negative supercoiling of plasmid DNA in *Escherichia coli* exposed to cold shock. *Molecular microbiology* 23, 381-386.

Moll, I., Grill, S., Grundling, A. & Blasi, U. (2002) Effects of ribosomal proteins S1, S2 and the DeaD/CsdA DEAD-box helicase on translation of leaderless and canonical mRNAs in *Escherichia coli*. *Molecular microbiology* 44, 1387-1396.

Moll, I., Huber, M., Grill, S., Sairafi, P., Mueller, F., Brimacombe, R., Londei, P. & Blasi, U. (2001) Evidence against an Interaction between the mRNA downstream box and 16S rRNA in translation initiation. *Journal of bacteriology* 183, 3499-3505.

Moser, M.J., Holley, W.R., Chatterjee, A. & Mian, I.S. (1997) The proofreading domain of *Escherichia coli* DNA polymerase I and other DNA and/or RNA exonuclease domains. *Nucleic ccids res* 25, 5110-5118.

Murata, N. & Wada, H. (1995) Acyl-lipid desaturases and their importance in the tolerance and acclimatization to cold of cyanobacteria. *The Biochemical journal* 308 (Pt 1), 1-8.

Nakaminami, K., Karlson, D.T. & Imai, R. (2006) Functional conservation of cold shock domains in bacteria and higher plants. *Proceedings of the National Academy of Sciences of the United States of America* 103, 10122-10127.

Narberhaus, F., Waldminghaus, T. & Chowdhury, S. (2006) RNA thermometers. *FEMS microbiology reviews* 30, 3-16.

Nishi, K., Morel-Deville, F., Hershey, J.W., Leighton, T. & Schnier, J. (1988) An eIF-4A-like protein is a suppressor of an *Escherichia coli* mutant defective in 50S ribosomal subunit assembly. *Nature* 336, 496-498.

Nishiyama, S.I., Umemura, T., Nara, T., Homma, M. & Kawagishi, I. (1999) Conversion of a bacterial warm sensor to a cold sensor by methylation of a single residue in the presence of an attractant. *Molecular microbiology* 32, 357-365.

Peil, L., Virumae, K. & Remme, J. (2008) Ribosome assembly in *Escherichia coli* strains lacking the RNA helicase DeaD/CsdA or DbpA. *The FEBS journal* 275, 3772-3782.

Phadtare, S. (2011) Unwinding activity of cold shock proteins and RNA metabolism. *RNA biology* 8, 394-397.

Phadtare, S. (2012) *Escherichia coli* cold-shock gene profiles in response to over-expression/deletion of CsdA, RNase R and PNPase and relevance to low-temperature RNA metabolism. *Genes to Cells* 17, 850-874.

Phadtare, S. & Inouye, M. (2004) Genome-wide transcriptional analysis of the cold shock response in wild-type and cold-sensitive, quadruple-*csp*-deletion strains of *Escherichia coli*. *Journal of bacteriology* 186, 7007-7014.

Phadtare, S., Inouye, M. & Severinov, K. (2002a) The nucleic acid melting activity of *Escherichia coli* CspE is critical for transcription antitermination and cold acclimation of cells. *The Journal of biological chemistry* 277, 7239-7245.

Phadtare, S., Inouye, M. & Severinov, K. (2004) The mechanism of nucleic acid melting by a CspA family protein. *Journal of molecular biology* 337, 147-155.

Phadtare, S. & Severinov, K. (2005) Nucleic acid melting by *Escherichia coli* CspE. *Nucleic acids research* 33, 5583-5590.

Phadtare, S. & Severinov, K. (2010) RNA remodeling and gene regulation by cold shock proteins. *RNA biology* 7, 788-795.

Phadtare, S., Tyagi, S., Inouye, M. & Severinov, K. (2002b) Three amino acids in *Escherichia coli* CspE surface-exposed aromatic patch are critical for nucleic acid melting activity leading to transcription antitermination and cold acclimation of cells. *The Journal of biological chemistry* 277, 46706-46711.

Porankiewicz, J. & Clarke, A.K. (1997) Induction of the heat shock protein ClpB affects cold acclimation in the cyanobacterium *Synechococcus* sp. strain PCC 7942. *Journal of bacteriology* 179, 5111-5117.

Proux, F., Dreyfus, M. & Iost, I. (2011) Identification of the sites of action of SrmB, a DEAD-box RNA helicase involved in *Escherichia coli* ribosome assembly. *Molecular microbiology* 82, 300-311.

Prud'homme-Genereux, A., Beran, R.K., Iost, I., Ramey, C.S., Mackie, G.A. & Simons, R.W. (2004) Physical and functional interactions among RNase E, polynucleotide phosphorylase and the cold-shock protein, CsdA: evidence for a 'cold shock degradosome'. *Molecular microbiology* 54, 1409-1421.

Qing, G., Xia, B. & Inouye, M. (2004) Enhancement of Translation Initiation by A/T-Rich Sequences Downstream of the Initiation Codon in *Escherichia coli*. *Journal of molecular microbiology biotechnology* 6, 133-144.

Rajkowitsch, L., Chen, D., Stampfl, S., Semrad, K., Waldsich, C., Mayer, O., Jantsch, M.F., Konrat, R., Blasi, U. & Schroeder, R. (2007) RNA chaperones, RNA annealers and RNA helicases. *RNA biology* 4, 118-130.

Remme, J. (2011) How to avoid undesirable interactions during ribosome assembly? *Molecular microbiology* 82, 269-271.

Runyon, G.T., Bear, D.G. & Lohman, T.M. (1990) *Escherichia coli* helicase II (UvrD) protein initiates DNA unwinding at nicks and blunt ends. *Proceedings of the National Academy of Sciences of the United States of America* 87, 6383-6387.

Shajani, Z., Sykes, M.T. & Williamson, J.R. (2011) Assembly of bacterial ribosomes. *Annual review of biochemistry* 80, 501-526.

Sinensky, M. (1974) Homeoviscous adaptation--a homeostatic process that regulates the viscosity of membrane lipids in *Escherichia coli*. *Proceedings of the National Academy of Sciences of the United States of America* 71, 522-525.

Storz, G. (1999) An RNA thermometer. *Genes & development* 13, 633-636.

Tanner, N.K., Cordin, O., Banroques, J., Doere, M. & Linder, P. (2003) The Q motif: a newly identified motif in DEAD box helicases may regulate ATP binding and hydrolysis. *Molecular cell* 11, 127-138.

Toone, W.M., Rudd, K.E. & Friesen, J.D. (1991) *deaD*, a new *Escherichia coli* gene encoding a presumed ATP-dependent RNA helicase, can suppress a mutation in *rpsB*, the gene encoding ribosomal protein S2. *Journal of bacteriology* 173, 3291-3302.

Trubetskoy, D., Proux, F., Allemand, F., Dreyfus, M. & Iost, I. (2009) SrmB, a DEAD-box helicase involved in *Escherichia coli* ribosome assembly, is specifically targeted to 23S rRNA in vivo. *Nucleic acids research* 37, 6540-6549.

Turner, A.M., Love, C.F., Alexander, R.W. & Jones, P.G. (2007) Mutational Analysis of *Escherichia coli* DEAD-box Protein CsdA. *Journal of bacteriology* 189, 2769-76.

Tuteja, N. & Tuteja, R. (2004) Prokaryotic and eukaryotic DNA helicases - Essential molecular motor proteins for cellular machinery. *European journal of biochemistry* 271, 1835-1848.

VanBogelen, R.A. & Neidhardt, F.C. (1990) Ribosomes as sensors of heat and cold shock in *Escherichia coli*. *Proceedings of the National Academy of Sciences of the United States of America* 87, 5589-5593.

Voloshin, O.N. & Camerini-Otero, R.D. (2007) The DinG protein from *Escherichia coli* is a structure-specific helicase. The *Journal of biological chemistry* 282, 18437-18447.

Wang, J.Y. & Syvanen, M. (1992) DNA twist as a transcriptional sensor for environmental changes. *Molecular microbiology* 6, 1861-1866.

Xia, B., Etchegaray, J.P. & Inouye, M. (2001a) Nonsense mutations in *cspA* cause ribosome trapping leading to complete growth inhibition and cell death at low temperature in *Escherichia coli*. *Journal of Biology and Chemistry* 276, 35581-35588.

Xia, B., Ke, H. & Inouye, M. (2001b) Acquirement of cold sensitivity by quadruple deletion of the cspA family and its suppression by PNPase S1 domain in *Escherichia coli*. *Molecular microbiology* 40, 179-188.

Xia, B., Ke, H., Shinde, U. & Inouye, M. (2003) The role of RbfA in 16S rRNA processing and cell growth at low temperature in *Escherichia coli. Journal of molecular biology* 332, 575-584.

Yamanaka, K. & Inouye, M. (2001) Selective mRNA degradation by polynucleotide phosphorylase in cold shock adaptation in *Escherichia coli. Journal of bacteriology* 183, 2808-2816.

Yamanaka, K., Mitta, M. & Inouye, M. (1999) Mutation analysis of the 5 ' untranslated region of the cold shock *cspA* mRNA of *Escherichia coli. Journal of bacteriology* 181, 6284-6291.

In: Ribosomes
Editors: Zhou Lin and Wang Liu

ISBN: 978-1-62417-698-2

Chapter 6

Structure, Function and Potential Biomedical Use of Ribosome-Inactivating Proteins

***Hong Xu*[*1] *and Wang-Yi Liu*[2]**
[1]Institute of Life Science, Nanchang University, Nanchang, China
[2]Institute of Biochemistry and Cell Biology,
Chinese Academy of Sciences, Shanghai, China

Abstract

Ribosome is crucial for protein synthesis in cells. Ribosome-inactivation proteins (RIPs) are a large family of unique enzymes that inhibit protein synthesis by site-specifically and irreversibly modifying the ribosomal RNA. According to their different enzymatic activity, RIPs can be classified into ribonuclease RIPs and N-glycosidase RIPs. RIPs have drawn the attention of many researchers due to their potential use as antivirus and anticancer reagents and as other biological and biomedical tools. In this chapter, the progress in the study of RIPs and the problems encountered in biological and biomedical research are discussed.

Introduction

The ribosome is a critical molecular machine for the translation of messenger RNA (mRNA) into peptide chains in nearly all living cells. Ribosomal RNA (rRNA) is key component of ribosome and served as ribozyme in protein synthesis [1]. There are many protein synthesis inhibitors targeting ribosomal RNA, including a lot antibiotics and ribosome-inactivating proteins (RIPs) [2, 3]. RIPs are a large family of naturally-occurring

* Corresponding email: hongxucpu@hotmail.com.

proteins that arrest protein synthesis by site-specifically and irreversibly modifying ribosomal RNA (rRNA). According to their enzymatic activity, RIPs are classified into two families, ribonuclease RIPs and N-glycosidase RIPs, which are represented by alpha-sarcin and ricin, respectively. We will discuss the potential anti-cancer and anti-viral activity and other biological and biomedical use of RIPs in this chapter. We do not intend to review the thorough literatures and only pick some examples to support our view.

1. Classification of RIPs

RIPs inhibit protein synthesis by catalyzing site-specific and irreversible modification of rRNA in intact ribosomes. Based on the different enzymatic activity, we propose that RIPs should be divided into two sub-families, ribonuclease RIPs and N-glycosidase RIPs.

Ribonuclease RIPs

Alpha-sarcin is a cytotoxic protein with anti-tumor activity isolated from fungi [4]. It inhibits protein synthesis by site-specifically cleaving intact ribosome at a single phosphodiester bond located, in rat ribosome, between G4325 and A4326 of 28S rRNA [5] (Figure 1). The single cleavage by alpha-sarcin releases an oligonucleotide fragment from the 3' end of 28S rRNA, which is termed alpha-fragment. The ribonuclease activity of alpha-sarcin is unusually specific when the substrate is intact ribosome or the 60S subunit. In contrast, the 40S subunit is not affected. Alpha-sarcin hydrolyzes only one out of more than 7000 phosphodiester bonds in the rRNAs. The break of the single phosphodiester bond prevents both the EF-1 (elongation factor 1)-catalyzed binding of aminoacyl-tRNA and the GTP-dependent binding of EF-2 to ribosomes [6]. Besides alpha-sarcin, similar ribonuclease RIPs have also been characterized, such as restrictocin and mitogillin [7]. They inhibit protein synthesis in the same way as alpha-sarcin does [8].

Besides the sarcin family ribonuclease RIPs, a novel ribonuclease RIP was isolated in our laboratory from the seeds of *Biota orientalis* and it was named *Biota orientalis RNase* [9]. Different from alphar-sarcin, the phosphodiester bond hydrolyzed by *Biota orientalis* RNase is located between C4453 and A4454 of rat 28S rRNA, which is 128 nucleic acids downstream of alpha-sarcin cleavage site [9] (Figure 1). Similar to alpha-sarcin, *Biota orientalis* RNase blocks protein synthesis by hydrolyzing a single phosphodiester bond of the ribosome. However, unlike alpha-sarcin, *Biota orientalis* RNase only interferes with EF-1-dependent binding of aminoacyl-tRNA to ribosome without affecting EF-2 binding [9].

Another type of ribonuclease RIP is *Colicin E3*, which is synthesized and secreted by certain strains of bacteria and is highly toxic to many strains of E. coli [10]. Totally different from alpha-sarcin, Colicin E3 only cleaves prokaryotic ribosomes, instead of eukaryotic ribosomes, at a single phosphodiester bond of the E. coli small subunit 16S rRNA between A1493 and G1494 [11].

N-Glycosidase RIPs

Distinguished from the above mentioned ribonuclease RIPs, the rest of RIPs, including ricin, abrin, pokeweed antivirus protein (PAP), trichosanthin and *et al.*, possess N-glycosidase activity. The N-glycosidase RIPs block protein synthesis by site-specifically cleaving a single N-C glycosidic bond between an adenine and the nucleotide RNA and thereby removing the adenine base (depurination) from the sugar-phosphate backbone of 28S rRNA [12]. The depurinated (or abasic) site becomes very unstable and beta-elimination reaction may occur under the treatment with mild alkaline or aniline at acidic pH. The nucleotide attacked by N-glycosidase RIPs (A4324 in 28S rRNA of rat liver ribosome) is located immediately next to the targeted site by alpha-sarcin (between G4325 and A4326 of rat 28S rRNA) [12]. The targeted sites by ricin and sarcin are located in the loop of a highly conserved stem-loop structure, which is accordingly named sarcin/ricin (S/R) domain [13]. The S/R domain is crucial for the protein synthesis function of ribosome in that it is involved in EF-1 dependent binding of aminoacyl-tRNA and EF-2-catalyzed GTP hydrolysis and translocation [14].

Based on the nature and number of subunits, N-glycosidase RIPs can be sub-divided into two major classes, type I and type II [15-22]. Type I N-glycosidase RIPs are single peptide chains containing an enzymatic domain. Type II N-glycosidase RIPs are composed of two chains linked by a disulfide bond, one enzymatic A-chain similar to type I RIP and one B-chain with lectin property that can bind carbohydrate and facilitates the endocytosis of type II RIPs. N-glycosidase RIPs were mainly discovered in high plants [23] and a few were found in alga [24], fungi and bacteria [18, 21, 25]. Most of the identified N-glycosidase RIPs belong to type I family and the number of type II ones is relatively small. The molecular weight of type I N-glycosidase RIPs is usually around 30kD, however, some type I RIPs are as small as 10kD [15, 16]. The primary sequences of type I N-glycosidase RIPs are remarkably similar to each other, especially in the enzymatic domains. There is also high similarity between type I RIP and the A-chain of type II RIPs. The B-chains of type II RIPs are also strikingly homologous with each other.

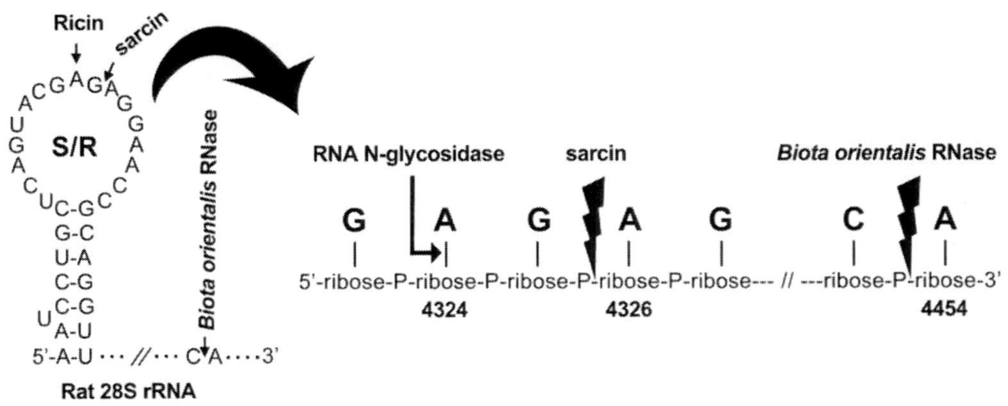

Figure 1. The cleavage sites by N-glycosidase RIPs and ribonuclease RIPs on rat 28S rRNA [5, 9, 12, 13]. The N-glycosidase RIPs cleave the N-C glycosidic bond at A4324 of rat 28S rRNA while the ribonuclease RIPs sarcin and *Biota orientalis* RNase hydrolyze indicated phosphodiester bonds. The cleavage sites by N-glycosidase and sarcin are immediately next to each other while those by sarcin and *Biota orientalis* RNase (A4326 and A4454) are 128 nucleotides away.

2. 'Extra' Enzymatic Activities of RIPs

Almost all RIPs, including both ribonuclease RIPs and N-glycosidase RIPs, can hydrolyze intact ribosomes in a dramatically high site-selectivity manner. Under certain conditions, they only attack the ribosome at a single nucleotide out of thousands of nucleic acids in rRNAs. However, the selectivity of RIPs varies dependent upon not only the enzymes but also the substrates. For example, when naked total rRNA or 28S rRNA is the substrate, alpha-sarcin exhibits endonuclease activity and causes extensive hydrolysis of the nucleic acids [26].

Similar to ribonuclease RIPs, although N-glycosidase RIPs are characterized by the site-specific depurination activity on ribosomes, the selectivity might change under certain conditions. For example, some N-glycosidase RIPs have been observed to remove more than one adenine per ribosome [27]. It has also been reported that N-glycosidase RIPs can also depurinate single-stranded DNA, double-stranded DNA, mRNA, tRNA, poly(A), and even viral RNA other than ribosomal RNA [28-34]. Accodingly, the denomination polynucleotide:adenosine glycosidase was proposed [30]. Furthermore, some N-glycosidase RIPs were shown to depurinate poly(ADP-ribosyl)ated poly(ADP-ribose) polymerase (PARP) [35, 36], an enzyme involved in DNA repair. Supercoiled double-stranded DNAs have been reported to be the substrate of N-glycosidase RIPs and they can be deadenylated and converted into nicked and linear forms [37, 38].

The 'extra' enzymatic activities of RIPs are still ambiguous and remain open to further investigation. The multiple biological functions of RIPs, such as cytotoxicity, anti-virus and anti-tumor activities, might be at least partly explained by these 'extra' enzymatic activities.

3. Antiviral Activity of RIPs

Many studies have demonstrated that RIPs possess antiviral activities against plant, fungal and animal viruses [15, 39]. The activity against animal virus is especially intriguing since it might provide therapy for some human diseases caused by viruses. RIPs exhibit broad-spectrum antiviral activity and they are potential inhibitors of many viruses such as HIV (human immunodeficiency virus), HBV (hepatitis B virus) and HSV (herpes simplex virus).

Trichosanthin

So far, most studied antiviral RIPs are type I N-glycosidase RIPs. Trichosanthin is an N-glycosidase purified from traditional Chinese medicinal herb Tian-Hua-Fen, an abortifacient reagent obtained from the root tubers of *Trichosanthes kirilowii* [40, 41]. GLQ223 is a highly purified, formulated preparation of trichosanthin and it was found to dose-dependently inhibit replication of HIV-1 in both acutely infected T-lymphoblastoid cells and chronically infected monocyte/macrophages *in vitro* [42]. The anti-viral activity is highly selective since that substantial arresting of viral replication was observed at the concentrations of GLQ223 that exhibited little inhibition of parallel unaffected cultures. Virus replication was dramatically blocked after only a single dose of GLQ223 in the cultures of the blood samples freshly drawn from HIV-1 affected patients [42]. The antiviral activity of trichosanthin is correlated with its ribosome inactivating activity, however, ribosome inhibition alone might not be adequate to explain the antiviral activity [43, 44]. Further study suggested that trichosanthin blocks HIV-1 replication through depurinating the long terminal repeats of HIV-1 DNA and preventing the integration of HIV-1 DNA into cellular genome [45]. The anti-HIV activity of trichosanthin is also attributed to the enhanced apoptosis of infected cells [46]. The replication of another virus, herpes simplex virus-1 (HSV-1), can also be inhibited by trichosanthin [47, 48]. The inhibition was proposed to be partly due to the induction of apoptosis in the HSV-1 infected cells [49, 50].

Pokeweed Antiviral Protein

Pokeweed antiviral protein (PAP) is an N-glycosidase RIP isolated from *Phytolacca Americana* [51, 52]. PAP inhibits HIV-1 replication at doses that were not toxic to uninfected peripheral blood mononuclear cells [52-54]. The anti-HIV activity of PAP is partly attributed to its ability to depurinate viral RNA and reduce the reverse transcriptase activity and the expression of p24 core protein in infected cells [55, 56]. Besides inhibiting HIV-1, PAP also blocks the replication of HBV, HSV and HTLV-1 (human T-cell leukemia virus) [57-59]. At the concentrations that PAP exhibits significant antiviral activities, no obvious cytotoxity is observed, which indicates that the antiviral activity of PAP is very specific to viruses [53, 60].

MAP30

Another most-studied antiviral RIP is MAP30, an antiviral protein isolated from *Momordica charantia* which can interfere with the infection and replication of HIV-1 [61-63]. MAP30 inhibits the expression of HIV-1 core protein p24 and the activity of viral reverse transcriptase [62]. It also prevents the integration of viral DNA into host genome by inhibiting the specific reactions catalyzed by the viral integrase [61]. MAP30 also shows potent inhibiting effects on HBV and HSV [64, 65].

Antiviral Activity *In Vivo*

The antiviral activity of RIPs has been mostly investigated in cell cultures and a few in animal models *in vivo* [66-68]. The results indicated that RIPs might be used as an alternative treatment for patients infected with viruses like HIV. Several clinical trials have been performed to evaluate the safety and effect of RIPs alone or in combination with other antiretroviral agents in virus-infected patients [42, 68-71]. Unfortunately, the results indicated that although RIPs, such as trichosanthin, were able to inhibit virus and increase $CD4^+$ and $CD8^+$ T cells in short term, they exhibited severe side effects [72-74]. Major side-effects were fatigue and myalgias. Reversible mild leucocytosis, elevations in serum transaminases and non-dose-related reversible mental status changes were also observed. Other side effects included fevers, flu-like syndrome and mild-moderate anaphylactic reactions [72-74]. Further studies are required to improve the selectivity of anti-viral RIPs, reduce the antigenic activity of the RIPs and prevent other side effects.

4. Induction of Apoptosis and Antitumor Activities of RIPs

Cytotoxicity of RIPs

It's well known that RIPs, especially some type II N-glycosidase RIPs, are highly toxic to cells, animals and humans. The cell toxicity of RIPs is usually well correlated with the toxicity to animals and humans [17, 18, 20, 21, 75-79]. Type II N-glycosidase RIPs are often more toxic than type I RIPs and this is mostly due to the existence of the B-chain in type II RIPs, which facilitates endocytosis of the A-chain into cells by binding carbohydrates on the cell surface [79-81]. However, not all type II RIPs exhibit high toxicity to cells and thus they are subdivided into two groups based on their cytotoxicity, toxic RIPs (such as ricin) and non-toxic RIPs (such as cinnamomin) [17, 80, 82]. Non-toxic type II RIPs are still toxic to cells, although much higher concentrations are required to induce the same level of cell death. For example, cinnamomin is about 138-fold less toxic to cells than ricin, although the N-glycosidase activity of the A-chains of the two RIPs is comparable. Followed subunit swap experiments demonstrated that the lower toxicity of cinnamomin resides in the B-chain. The cytotoxicity of the chimeric protein, reconstructed by combining the A-chain of cinnamomin with the B-chain of ricin, is comparable to that of native ricin. Meanwhile, the hybrid protein constructed from ricin A-chain and cinnamomin B-chain is significantly less toxic to cells than native ricin [80]. In domestic silkworm larvae, cinnamomin is remarkably less toxic than ricin, consistent with their differential *in vitro* cytotoxicity [83].

RIPs and Apoptosis

Initially, RIPs have been believed to induce cell necrosis in animal tissues. Later, it was revealed that ricin and abrin, two well-known type II N-glycosidase RIPs, induced apoptotic changes in lymphoid tissues and intestine of rats [84]. From that point on, many RIPs have been found to induce apoptosis [76, 77]. Some type I N-glycosidase RIPs, including pokeweed antiviral protein (PAP), trichosanthin, saporin, gelonin and *et al.*, have also been revealed to cause apoptotic cell death when they are delivered into cells [85-92].

Apoptosis and Enzymatic Activity of RIPs

It had been simply presumed that cells treated with RIPs died of inhibition of protein synthesis induced by the site-specific damage on ribosome. However, there have been increasing evidence contradicting with the idea and suggesting that protein synthesis inhibition is not the only mechanism of apoptosis [76-78]. For example, the isolated B-chain of ricin, which had no effect on cellular protein synthesis, induced apoptosis in a dose and time-dependent manner, possibly through its lectin activity, a mechanism apparently distinct from that of intact ricin [93]. In the screening of PAP with reduced cytotoxicity, several mutants were identified, including two mutants that were enzymatically active *in vitro* [94]. This suggested that the cytotoxicity of PAP is not solely due to its enzymatic activity against ribosome and protein synthesis. Different domains of PAP were revealed to be involved in the multiple signaling pathways of cytotoxicity [94]. Similarly, it was found that the catalytic activity of sarcin is not required for its cytotoxicity [95].

Signaling Pathways Involved in RIP-Induced Apoptosis

Several different signaling pathways have been proposed to be involved in the apoptotic cell death induced by RIPs [76, 77]. The classical one is the ribotoxic stress response induced by the N-glycosidase activity of RIPs on 28S rRNA, which might induce the activation of kinases involved in apoptosis, such as SAPK/JNK or p38 MAPK [96]. The hyperactivation of poly(ADP-ribose) polymerase (PARP) and thus the depletion of NAD^+ are also proposed to contribute to the apoptosis induction activity of RIPs [97]. As aforementioned, RIPs have been revealed to directly depurinate auto-modified PARP and release adenine from the ADP-ribosyl group [35, 36]. The depurination of PARP resulted in the unavailability of PARP for further ADP-ribosylation, which might cause the depletion of intracellular NAD^+ and thus trigger apoptosis [35]. Another signaling pathway involved in the apoptosis induced by RIPs might be the DNA damage caused by RIPs [98]. Besides depurinating 28S rRNA, many RIPs have been shown to depurinate DNA, mRNA and tRNA *in vitro* [28-34]. Furthermore, some RIPs were found to be able to cut supercoiled double stranded DNA into nicked and linear forms *in vitro* [37, 38]. Although contamination of nucleases has been reported in some RIP preparations [99], the idea that nuclear DNA and/or mitochondria DNA damage caused by RIPs induces apoptosis still could not be excluded. Other possible signaling pathways involved in RIP-induced apoptosis include stress induced mitochondria pathway [100], downregulation of the anti-apoptotic factors and inhibition of antioxidant proteins [77].

Anticancer and RIPs

In spite of the fact that the mechanism of RIP-induced cytotoxicity is not clear yet, RIPs are still promising candidates to be used as therapy for human disease, especially cancer. Since the 1920s, the extract of European mistletoe (*Viscum album*) has been used for treating cancer, especially in Europe. In Africa, the extract from *X. Americana* has long been used as a traditional medicine in cancer patients. It was later found that these extracts contain type II N-glycosidase RIPs, viscumin and riproximin, respectively [101, 102]. RIPs are among the most powerful killing agents and, theoretically, cytosolic entry of one single molecule is enough to kill the cell. The extremely high toxicity and selectivity of RIPs has stimulated the hope of developing anticancer reagents with them. Usually the single chain type I N-glycosidase RIPs or the A-chain of type II RIPs, are chemically linked to or genetically recombined with carrier molecules, such as antibodies, hormones and growth factors that can direct the toxic RIPs into certain cells and tissues in a selective manner [103]. Many preclinical and clinical investigations on the effect of conjugated RIPs have been performed in the past years.

Unfortunately, the RIP-derived conjugates have disadvantages and side effects, such as being immunogenic and causing vascular leak syndrome [103]. Further studies are required in order to overcome these side effects and improve their selectivity.

5. Other Biomedical Use of RIPs

Besides the use of RIPs as anticancer and antiviral reagents, RIPs have also been used to selectively lesion specific neurons, which method is called 'molecular neurosurgery' [104]. RIPs are usually conjugated with specific surface markers, such as neurotransmitter receptor, neurotransmitter transporter, growth factor receptors, the G-protein coupled receptors for neuropeptides and *et al.*, to selectively deliver RIPs into certain neurons and then kill them. The molecular neurosurgery method has been widely used to dissect neural circuits involved in different functions of the nervous system. The conjugated RIPs can be used as a pain killer to ablate the neurons involved in chronic pain with high selectivity [105]. Similarly, they can be used to specifically disable muscles and may be useful in the treatment of strabismus or spasm [106, 107].

Type II RIPs are highly toxic and there are accidents caused by type II RIPs every year, such as ingestion of castor bean, which contains highly toxic RIP ricin. Ricin has been concerned that it could be used as a bioweapon of mass destruction or to specifically target an enemy. Ricin might also be used as terrorist attack due to its wide availability and easy preparation [108].

Conclusion

Our knowledge of RIPs has dramatically increased in the past years. Their potential use as antivirus and anticancer reagents has stimulated researchers to study their structure, activity and function. Undoubtedly, many aspects of RIPs still require deep investigations. For example, the mechanism of cell entry and toxicity of RIPs still remains open. Clinical trials have illustrated that RIPs might cause many side effects. Further studies are needed to refine and improve conjugates using RIPs as a component to target specific cells, such as tumor cells, neurons and virus-infected cells.

References

[1] Simonovic M, Steitz TA. A structural view on the mechanism of the ribosome-catalyzed peptide bond formation. *Biochim Biophys Acta*. 2009 Sep-Oct;1789(9-10):612-23.

[2] Yonath A. Antibiotics targeting ribosomes: resistance, selectivity, synergism and cellular regulation. *Annu Rev Biochem*. 2005;74:649-79.

[3] Stirpe F, Barbieri L, Battelli MG, Soria M, Lappi DA. Ribosome-inactivating proteins from plants: present status and future prospects. *Biotechnology (N Y)*. 1992 Apr;10(4):405-12.

[4] Jennings JC, Olson BH, Roga V, Junek AJ, Schuurmans DM. Alpha Sarcin, a New Antitumor Agent. Ii. Fermentation and Antitumor Spectrum. *Appl Microbiol.* 1965 May;13:322-6.

[5] Endo Y, Wool IG. The site of action of alpha-sarcin on eukaryotic ribosomes. The sequence at the alpha-sarcin cleavage site in 28 S ribosomal ribonucleic acid. *J Biol Chem.* 1982 Aug 10;257(15):9054-60.

[6] Brigotti M, Rambelli F, Zamboni M, Montanaro L, Sperti S. Effect of alpha-sarcin and ribosome-inactivating proteins on the interaction of elongation factors with ribosomes. *Biochem J.* 1989 Feb 1;257(3):723-7.

[7] Rodriguez R, Lopez-Otin C, Barber D, Fernandez-Luna JL, Gonzalez G, Mendez E. Amino acid sequence homologies in alfa-sarcin, restrictocin and mitogillin. *Biochem Biophys Res Commun.* 1982 Sep 16;108(1):315-21.

[8] Fando JL, Alaba I, Escarmis C, Fernandez-Luna JL, Mendez E, Salinas M. The mode of action of restrictocin and mitogillin on eukaryotic ribosomes. Inhibition of brain protein synthesis, cleavage and sequence of the ribosomal RNA fragment. *Eur J Biochem.* 1985 May 15;149(1):29-34.

[9] Xu H, He WJ, Liu WY. A novel ribotoxin with ribonuclease activity that specifically cleaves a single phosphodiester bond in rat 28S ribosomal RNA and inactivates ribosome. *Arch Biochem Biophys.* 2004 Jul 1;427(1):30-40.

[10] Nomura M. Colicins and related bacteriocins. *Annu Rev Microbiol.* 1967;21:257-84.

[11] Bowman CM, Dahlberg JE, Ikemura T, Konisky J, Nomura M. Specific inactivation of 16S ribosomal RNA induced by colicin E3 in vivo. *Proc Natl Acad Sci U S A.* 1971 May;68(5):964-8.

[12] Endo Y, Mitsui K, Motizuki M, Tsurugi K. The mechanism of action of ricin and related toxic lectins on eukaryotic ribosomes. The site and the characteristics of the modification in 28 S ribosomal RNA caused by the toxins. *J Biol Chem.* 1987 Apr 25;262(12):5908-12.

[13] Endo Y, Gluck A, Wool IG. Ribosomal RNA identity elements for ricin A-chain recognition and catalysis. *J Mol Biol.* 1991 Sep 5;221(1):193-207.

[14] Gluck A, Endo Y, Wool IG. The ribosomal RNA identity elements for ricin and for alpha-sarcin: mutations in the putative CG pair that closes a GAGA tetraloop. *Nucleic Acids Res.* 1994 Feb 11;22(3):321-4.

[15] Puri M, Kaur I, Perugini MA, Gupta RC. Ribosome-inactivating proteins: current status and biomedical applications. *Drug Discov Today.* 2012 Mar 30.

[16] Ng TB, Wong JH, Wang H. Recent progress in research on ribosome inactivating proteins. *Curr Protein Pept Sci.* 2010 Feb;11(1):37-53.

[17] Stirpe F, Battelli MG. Ribosome-inactivating proteins: progress and problems. *Cell Mol Life Sci.* 2006 Aug;63(16):1850-66.

[18] Stirpe F. Ribosome-inactivating proteins. *Toxicon.* 2004 Sep 15;44(4):371-83.

[19] Robertus JD, Monzingo AF. The structure of ribosome inactivating proteins. *Mini Rev Med Chem.* 2004 Jun;4(5):477-86.

[20] Peumans WJ, Hao Q, Van Damme EJ. Ribosome-inactivating proteins from plants: more than RNA N-glycosidases? *FASEB J.* 2001 Jul;15(9):1493-506.

[21] Nielsen K, Boston RS. RIBOSOME-INACTIVATING PROTEINS: A Plant Perspective. *Annu Rev Plant Physiol Plant Mol Biol.* 2001 Jun;52:785-816.

[22] Singh RC, Singh V. Current status of ribosome inactivating proteins. *Indian J Biochem Biophys*. 2000 Feb;37(1):1-5.

[23] Barbieri L, Battelli MG, Stirpe F. Ribosome-inactivating proteins from plants. *Biochim Biophys Acta*. 1993 Dec 21;1154(3-4):237-82.

[24] Liu RS, Yang JH, Liu WY. Isolation and enzymatic characterization of lamjapin, the first ribosome-inactivating protein from cryptogamic algal plant (Laminaria japonica A). *Eur J Biochem*. 2002 Oct;269(19):4746-52.

[25] Girbes T, Ferreras JM, Arias FJ, Stirpe F. Description, distribution, activity and phylogenetic relationship of ribosome-inactivating proteins in plants, fungi and bacteria. *Mini Rev Med Chem*. 2004 Jun;4(5):461-76.

[26] Endo Y, Huber PW, Wool IG. The ribonuclease activity of the cytotoxin alpha-sarcin. The characteristics of the enzymatic activity of alpha-sarcin with ribosomes and ribonucleic acids as substrates. *J Biol Chem*. 1983 Feb 25;258(4):2662-7.

[27] Barbieri L, Ferreras JM, Barraco A, Ricci P, Stirpe F. Some ribosome-inactivating proteins depurinate ribosomal RNA at multiple sites. *Biochem J*. 1992 Aug 15;286 (Pt 1):1-4.

[28] Bagga S, Seth D, Batra JK. The cytotoxic activity of ribosome-inactivating protein saporin-6 is attributed to its rRNA N-glycosidase and internucleosomal DNA fragmentation activities. *J Biol Chem*. 2003 Feb 14;278(7):4813-20.

[29] Hudak KA, Wang P, Tumer NE. A novel mechanism for inhibition of translation by pokeweed antiviral protein: depurination of the capped RNA template. *RNA*. 2000 Mar;6(3):369-80.

[30] Barbieri L, Valbonesi P, Bonora E, Gorini P, Bolognesi A, Stirpe F. Polynucleotide:adenosine glycosidase activity of ribosome-inactivating proteins: effect on DNA, RNA and poly(A). *Nucleic Acids Res*. 1997 Feb 1;25(3):518-22.

[31] Nicolas E, Beggs JM, Haltiwanger BM, Taraschi TF. A new class of DNA glycosylase/apurinic/apyrimidinic lyases that act on specific adenines in single-stranded DNA. *J Biol Chem*. 1998 Jul 3;273(27):17216-20.

[32] Barbieri L, Gorini P, Valbonesi P, Castiglioni P, Stirpe F. Unexpected activity of saporins. *Nature*. 1994 Dec 15;372(6507):624.

[33] Barbieri L, Valbonesi P, Brigotti M, Montanaro L, Stirpe F, Sperti S. Shiga-like toxin I is a polynucleotide:adenosine glycosidase. *Mol Microbiol*. 1998 Jul;29(2):661-2.

[34] Brigotti M, Barbieri L, Valbonesi P, Stirpe F, Montanaro L, Sperti S. A rapid and sensitive method to measure the enzymatic activity of ribosome-inactivating proteins. *Nucleic Acids Res*. 1998 Sep 15;26(18):4306-7.

[35] Barbieri L, Brigotti M, Perocco P, Carnicelli D, Ciani M, Mercatali L, et al. Ribosome-inactivating proteins depurinate poly(ADP-ribosyl)ated poly(ADP-ribose) polymerase and have transforming activity for 3T3 fibroblasts. *FEBS Lett*. 2003 Mar 13;538(1-3):178-82.

[36] Sestili P, Alfieri R, Carnicelli D, Martinelli C, Barbieri L, Stirpe F, et al. Shiga toxin 1 and ricin inhibit the repair of H2O2-induced DNA single strand breaks in cultured mammalian cells. *DNA Repair (Amst)*. 2005 Feb 3;4(2):271-7.

[37] Wang P, Tumer NE. Pokeweed antiviral protein cleaves double-stranded supercoiled DNA using the same active site required to depurinate rRNA. *Nucleic Acids Res*. 1999 Apr 15;27(8):1900-5.

[38] He WJ, Liu WY. Both N- and C-terminal regions are essential for cinnamomin A-chain to deadenylate ribosomal RNA and supercoiled double-stranded DNA. *Biochem J.* 2004 Jan 1;377(Pt 1):17-23.

[39] Kaur I, Gupta RC, Puri M. Ribosome inactivating proteins from plants inhibiting viruses. *Virol Sin.* 2011 Dec;26(6):357-65.

[40] Zhang JS, Liu WY. The mechanism of action of trichosanthin on eukaryotic ribosomes--RNA N-glycosidase activity of the cytotoxin. *Nucleic Acids Res.* 1992 Mar 25;20(6):1271-5.

[41] Fang EF, Ng TB, Shaw PC, Wong RN. Recent progress in medicinal investigations on trichosanthin and other ribosome inactivating proteins from the plant genus Trichosanthes. *Curr Med Chem.* 2011;18(28):4410-7.

[42] McGrath MS, Hwang KM, Caldwell SE, Gaston I, Luk KC, Wu P, et al. GLQ223: an inhibitor of human immunodeficiency virus replication in acutely and chronically infected cells of lymphocyte and mononuclear phagocyte lineage. *Proc Natl Acad Sci U S A.* 1989 Apr;86(8):2844-8.

[43] Wang JH, Nie HL, Huang H, Tam SC, Zheng YT. Independency of anti-HIV-1 activity from ribosome-inactivating activity of trichosanthin. *Biochem Biophys Res Commun.* 2003 Feb 28;302(1):89-94.

[44] Wang JH, Nie HL, Tam SC, Huang H, Zheng YT. Anti-HIV-1 property of trichosanthin correlates with its ribosome inactivating activity. *FEBS Lett.* 2002 Nov 6;531(2):295-8.

[45] Zhao WL, Feng D, Wu J, Sui SF. Trichosanthin inhibits integration of human immunodeficiency virus type 1 through depurinating the long-terminal repeats. *Mol Biol Rep.* 2010 Apr;37(4):2093-8.

[46] Wang YY, Ouyang DY, Huang H, Chan H, Tam SC, Zheng YT. Enhanced apoptotic action of trichosanthin in HIV-1 infected cells. *Biochem Biophys Res Commun.* 2005 Jun 17;331(4):1075-80.

[47] He DX, Tam SC. Trichosanthin affects HSV-1 replication in Hep-2 cells. *Biochem Biophys Res Commun.* 2010 Nov 26;402(4):670-5.

[48] Zheng YT, Chan WL, Chan P, Huang H, Tam SC. Enhancement of the anti-herpetic effect of trichosanthin by acyclovir and interferon. *FEBS Lett.* 2001 May 11;496(2-3):139-42.

[49] He D, Zheng Y, Tam S. The anti-herpetic activity of trichosanthin via the nuclear factor-kappaB and p53 pathways. *Life Sci.* 2012 May 15;90(17-18):673-81.

[50] He D, Yau K, He X, Shi H, Zheng Y, Tam S. Conversion of trichosanthin-induced CD95 (Fas) type I into type II apoptotic signaling during Herpes simplex virus infection. *Mol Immunol.* 2011 Sep;48(15-16):2000-8.

[51] Wyatt SD, Shepherd RJ. Isolation and characterization of a virus inhibitor from Phytolacca americana. *Phytopathology.* 1969 Dec;59(12):1787-94.

[52] Irvin JD, Uckun FM. Pokeweed antiviral protein: ribosome inactivation and therapeutic applications. *Pharmacol Ther.* 1992;55(3):279-302.

[53] Zarling JM, Moran PA, Haffar O, Sias J, Richman DD, Spina CA, et al. Inhibition of HIV replication by pokeweed antiviral protein targeted to CD4+ cells by monoclonal antibodies. *Nature.* 1990 Sep 6;347(6288):92-5.

[54] Kim YW, Fung MS, Sun NC, Sun CR, Chang NT, Chang TW. Immunoconjugates that neutralize HIV virions kill T cells infected with diverse strains of HIV-1. *J Immunol.* 1990 Feb 15;144(4):1257-62.

[55] Rajamohan F, Venkatachalam TK, Irvin JD, Uckun FM. Pokeweed antiviral protein isoforms PAP-I, PAP-II, and PAP-III depurinate RNA of human immunodeficiency virus (HIV)-1. *Biochem Biophys Res Commun.* 1999 Jul 5;260(2):453-8.

[56] Olson MC, Ramakrishnan S, Anand R. Ribosomal inhibitory proteins from plants inhibit HIV-1 replication in acutely infected peripheral blood mononuclear cells. *AIDS Res Hum Retroviruses.* 1991 Dec;7(12):1025-30.

[57] He YW, Guo CX, Pan YF, Peng C, Weng ZH. Inhibition of hepatitis B virus replication by pokeweed antiviral protein in vitro. *World J Gastroenterol.* 2008 Mar 14;14(10):1592-7.

[58] Teltow GJ, Irvin JD, Aron GM. Inhibition of herpes simplex virus DNA synthesis by pokeweed antiviral protein. *Antimicrob Agents Chemother.* 1983 Mar;23(3):390-6.

[59] Mansouri S, Choudhary G, Sarzala PM, Ratner L, Hudak KA. Suppression of human T-cell leukemia virus I gene expression by pokeweed antiviral protein. *J Biol Chem.* 2009 Nov 6;284(45):31453-62.

[60] Uckun FM, Chelstrom LM, Tuel-Ahlgren L, Dibirdik I, Irvin JD, Langlie MC, et al. TXU (anti-CD7)-pokeweed antiviral protein as a potent inhibitor of human immunodeficiency virus. *Antimicrob Agents Chemother.* 1998 Feb;42(2):383-8.

[61] Lee-Huang S, Huang PL, Bourinbaiar AS, Chen HC, Kung HF. Inhibition of the integrase of human immunodeficiency virus (HIV) type 1 by anti-HIV plant proteins MAP30 and GAP31. *Proc Natl Acad Sci U S A.* 1995 Sep 12;92(19):8818-22.

[62] Lee-Huang S, Huang PL, Chen HC, Bourinbaiar A, Huang HI, Kung HF. Anti-HIV and anti-tumor activities of recombinant MAP30 from bitter melon. *Gene.* 1995 Aug 19;161(2):151-6.

[63] Lee-Huang S, Huang PL, Nara PL, Chen HC, Kung HF, Huang P, et al. MAP 30: a new inhibitor of HIV-1 infection and replication. *FEBS Lett.* 1990 Oct 15;272(1-2):12-8.

[64] Fan JM, Zhang Q, Xu J, Zhu S, Ke T, Gao de F, et al. Inhibition on Hepatitis B virus in vitro of recombinant MAP30 from bitter melon. *Mol Biol Rep.* 2009 Feb;36(2):381-8.

[65] Bourinbaiar AS, Lee-Huang S. The activity of plant-derived antiretroviral proteins MAP30 and GAP31 against herpes simplex virus in vitro. *Biochem Biophys Res Commun.* 1996 Feb 27;219(3):923-9.

[66] Uckun FM, Rajamohan F, Pendergrass S, Ozer Z, Waurzyniak B, Mao C. Structure-based design and engineering of a nontoxic recombinant pokeweed antiviral protein with potent anti-human immunodeficiency virus activity. *Antimicrob Agents Chemother.* 2003 Mar;47(3):1052-61.

[67] Uckun FM, Rustamova L, Vassilev AO, Tibbles HE, Petkevich AS. CNS activity of Pokeweed anti-viral protein (PAP) in mice infected with lymphocytic choriomeningitis virus (LCMV). *BMC Infect Dis.* 2005 Feb 22;5:9.

[68] Uckun FM, Bellomy K, O'Neill K, Messinger Y, Johnson T, Chen CL. Toxicity, biological activity, and pharmacokinetics of TXU (anti-CD7)-pokeweed antiviral protein in chimpanzees and adult patients infected with human immunodeficiency virus. *J Pharmacol Exp Ther.* 1999 Dec;291(3):1301-7.

[69] Kahn JO, Gorelick KJ, Gatti G, Arri CJ, Lifson JD, Gambertoglio JG, et al. Safety, activity, and pharmacokinetics of GLQ223 in patients with AIDS and AIDS-related complex. *Antimicrob Agents Chemother.* 1994 Feb;38(2):260-7.

[70] Garcia PA, Bredesen DE, Vinters HV, Graefin von Einsiedel R, Williams RL, Kahn JO, et al. Neurological reactions in HIV-infected patients treated with trichosanthin. *Neuropathol Appl Neurobiol.* 1993 Oct;19(5):402-5.

[71] Mayer RA, Sergios PA, Coonan K, O'Brien L. Trichosanthin treatment of HIV-induced immune dysregulation. *Eur J Clin Invest.* 1992 Feb;22(2):113-22.

[72] Byers VS, Levin AS, Malvino A, Waites L, Robins RA, Baldwin RW. A phase II study of effect of addition of trichosanthin to zidovudine in patients with HIV disease and failing antiretroviral agents. *AIDS Res Hum Retroviruses.* 1994 Apr;10(4):413-20.

[73] Kahn JO, Kaplan LD, Gambertoglio JG, Bredesen D, Arri CJ, Turin L, et al. The safety and pharmacokinetics of GLQ223 in subjects with AIDS and AIDS-related complex: a phase I study. *AIDS.* 1990 Dec;4(12):1197-204.

[74] Byers VS, Levin AS, Waites LA, Starrett BA, Mayer RA, Clegg JA, et al. A phase I/II study of trichosanthin treatment of HIV disease. *AIDS.* 1990 Dec;4(12):1189-96.

[75] Tesh VL. The induction of apoptosis by Shiga toxins and ricin. *Curr Top Microbiol Immunol.* 2012;357:137-78.

[76] Das MK, Sharma RS, Mishra V. Induction of apoptosis by ribosome inactivating proteins: importance of N-glycosidase activity. *Appl Biochem Biotechnol.* 2012 Mar;166(6):1552-61.

[77] Narayanan S, Surendranath K, Bora N, Surolia A, Karande AA. Ribosome inactivating proteins and apoptosis. *FEBS Lett.* 2005 Feb 28;579(6):1324-31.

[78] Battelli MG. Cytotoxicity and toxicity to animals and humans of ribosome-inactivating proteins. *Mini Rev Med Chem.* 2004 Jun;4(5):513-21.

[79] Lord MJ, Jolliffe NA, Marsden CJ, Pateman CS, Smith DC, Spooner RA, et al. Ricin. Mechanisms of cytotoxicity. *Toxicol Rev.* 2003;22(1):53-64.

[80] Wang BZ, Zou WG, Liu WY, Liu XY. The lower cytotoxicity of cinnamomin (a type II RIP) is due to its B-chain. *Arch Biochem Biophys.* 2006 Jul 1;451(1):91-6.

[81] Hartley MR, Lord JM. Cytotoxic ribosome-inactivating lectins from plants. *Biochim Biophys Acta.* 2004 Sep 1;1701(1-2):1-14.

[82] Xu H, Liu WY. Cinnamomin--a versatile type II ribosome-inactivating protein. *Acta Biochim Biophys Sin (Shanghai).* 2004 Mar;36(3):169-76.

[83] Wei GQ, Liu RS, Wang Q, Liu WY. Toxicity of two type II ribosome-inactivating proteins (cinnamomin and ricin) to domestic silkworm larvae. *Arch Insect Biochem Physiol.* 2004 Dec;57(4):160-5.

[84] Griffiths GD, Leek MD, Gee DJ. The toxic plant proteins ricin and abrin induce apoptotic changes in mammalian lymphoid tissues and intestine. *J Pathol.* 1987 Mar;151(3):221-9.

[85] Irvin JD, Aron GM. Chemical modifications of pokeweed antiviral protein: effects upon ribosome inactivation, antiviral activity and cytotoxicity. *FEBS Lett.* 1982 Nov 1;148(1):127-30.

[86] Leung KN, Yeung HW, Leung SO. The immunomodulatory and antitumor activities of trichosanthin-an abortifacient protein isolated from tian-hua-fen (Trichosanthes kirilowii). *Asian Pac J Allergy Immunol.* 1986 Dec;4(2):111-20.

[87] Thorpe PE, Brown AN, Bremner JA, Jr., Foxwell BM, Stirpe F. An immunotoxin composed of monoclonal anti-Thy 1.1 antibody and a ribosome-inactivating protein from Saponaria officinalis: potent antitumor effects in vitro and in vivo. *J Natl Cancer Inst.* 1985 Jul;75(1):151-9.

[88] Stirpe F, Olsnes S, Pihl A. Gelonin, a new inhibitor of protein synthesis, nontoxic to intact cells. Isolation, characterization, and preparation of cytotoxic complexes with concanavalin A. *J Biol Chem*. 1980 Jul 25;255(14):6947-53.

[89] Bolognesi A, Polito L, Tazzari PL, Lemoli RM, Lubelli C, Fogli M, et al. In vitro anti-tumour activity of anti-CD80 and anti-CD86 immunotoxins containing type 1 ribosome-inactivating proteins. *Br J Haematol*. 2000 Aug;110(2):351-61.

[90] Waddick KG, Myers DE, Gunther R, Chelstrom LM, Chandan-Langlie M, Irvin JD, et al. In vitro and in vivo antileukemic activity of B43-pokeweed antiviral protein against radiation-resistant human B-cell precursor leukemia cells. *Blood*. 1995 Dec 1;86(11):4228-33.

[91] Tang NL, Chan WL, Ke YO, Mak MK, Lai FM, Tam SC. Acute renal failure and proximal tubule lesions after trichosanthin injection in rats. *Exp Mol Pathol*. 1997 Apr;64(2):78-89.

[92] Bergamaschi G, Perfetti V, Tonon L, Novella A, Lucotti C, Danova M, et al. Saporin, a ribosome-inactivating protein used to prepare immunotoxins, induces cell death via apoptosis. *Br J Haematol*. 1996 Jun;93(4):789-94.

[93] Hasegawa N, Kimura Y, Oda T, Komatsu N, Muramatsu T. Isolated ricin B-chain-mediated apoptosis in U937 cells. *Biosci Biotechnol Biochem*. 2000 Jul;64(7):1422-9.

[94] Hur Y, Hwang DJ, Zoubenko O, Coetzer C, Uckun FM, Tumer NE. Isolation and characterization of pokeweed antiviral protein mutations in Saccharomyces cerevisiae: identification of residues important for toxicity. *Proc Natl Acad Sci U S A*. 1995 Aug 29;92(18):8448-52.

[95] Alford SC, Pearson JD, Carette A, Ingham RJ, Howard PL. Alpha-sarcin catalytic activity is not required for cytotoxicity. *BMC Biochem*. 2009;10:9.

[96] Iordanov MS, Pribnow D, Magun JL, Dinh TH, Pearson JA, Chen SL, et al. Ribotoxic stress response: activation of the stress-activated protein kinase JNK1 by inhibitors of the peptidyl transferase reaction and by sequence-specific RNA damage to the alpha-sarcin/ricin loop in the 28S rRNA. *Mol Cell Biol*. 1997 Jun;17(6):3373-81.

[97] Komatsu N, Nakagawa M, Oda T, Muramatsu T. Depletion of intracellular NAD(+) and ATP levels during ricin-induced apoptosis through the specific ribosomal inactivation results in the cytolysis of U937 cells. *J Biochem*. 2000 Sep;128(3):463-70.

[98] Brigotti M, Alfieri R, Sestili P, Bonelli M, Petronini PG, Guidarelli A, et al. Damage to nuclear DNA induced by Shiga toxin 1 and ricin in human endothelial cells. *FASEB J*. 2002 Mar;16(3):365-72.

[99] Barbieri L, Valbonesi P, Righi F, Zuccheri G, Monti F, Gorini P, et al. Polynucleotide:Adenosine glycosidase is the sole activity of ribosome-inactivating proteins on DNA. *J Biochem*. 2000 Nov;128(5):883-9.

[100] Fujii J, Matsui T, Heatherly DP, Schlegel KH, Lobo PI, Yutsudo T, et al. Rapid apoptosis induced by Shiga toxin in HeLa cells. *Infect Immun*. 2003 May;71(5):2724-35.

[101] Voss C, Eyol E, Frank M, von der Lieth CW, Berger MR. Identification and characterization of riproximin, a new type II ribosome-inactivating protein with antineoplastic activity from Ximenia americana. *FASEB J*. 2006 Jun;20(8):1194-6.

[102] Endo Y, Tsurugi K, Franz H. The site of action of the A-chain of mistletoe lectin I on eukaryotic ribosomes. The RNA N-glycosidase activity of the protein. *FEBS Lett*. 1988 Apr 25;231(2):378-80.

[103] de Virgilio M, Lombardi A, Caliandro R, Fabbrini MS. Ribosome-inactivating proteins: from plant defense to tumor attack. *Toxins (Basel)*. 2010 Nov;2(11):2699-737.

[104] Wiley RG, Kline IR. Neuronal lesioning with axonally transported toxins. *J Neurosci Methods*. 2000 Nov 15;103(1):73-82.

[105] Wiley RG. Substance P receptor-expressing dorsal horn neurons: lessons from the targeted cytotoxin, substance P-saporin. *Pain*. 2008 May;136(1-2):7-10.

[106] Christiansen SP, Sandnas A, Prill R, Youle RJ, McLoon LK. Acute effects of the skeletal muscle-specific immunotoxin ricin-mAb 35 on extraocular muscles of rabbits. *Invest Ophthalmol Vis Sci*. 2000 Oct;41(11):3402-9.

[107] Hott JS, Dalakas MC, Sung C, Hallett M, Youle RJ. Skeletal muscle-specific immunotoxin for the treatment of focal muscle spasm. *Neurology*. 1998 Feb;50(2):485-91.

[108] Schep LJ, Temple WA, Butt GA, Beasley MD. Ricin as a weapon of mass terror--separating fact from fiction. *Environ Int*. 2009 Nov;35(8):1267-71.

In: Ribosomes
Editors: Zhou Lin and Wang Liu

ISBN: 978-1-62417-698-2

Chapter 7

Coarse-Graining the Nano-Machine Ribosome to Elucidate Its Functional Dynamics

Ozge Kurkcuoglu[1] and Pemra Doruker[2,*]
[1]Department of Chemical Engineering, Istanbul Technical University, Maslak, Istanbul, Turkey
[2]Department of Chemical Engineering and Polymer Research Center, Bogazici University, Bebek, Istanbul, Turkey

Abstract

Coarse-grained computational techniques can provide information on conformational dynamics of the supramolecule ribosome, for which full-atom approaches become computationally demanding. Elastic network models and coarse-grained molecular dynamics simulations are the most commonly used methods for simple descriptions of the biomolecular structure. Usually, an amino acid or a nucleotide residue is represented by a single node, and simplified potential energy functions are used for interacting nodes. These approaches have been successful in reproducing the micro-to-milliseconds dynamics information, such as the experimentally observed ratchet-like rotation of the ribosomal subunits. Furthermore, they have provided insights on the structure-dynamics-function relationship of the ribosomal complex. This review summarizes various coarse-graining approaches on the nano-machine and recent studies that focus on its different functional sites.

Introduction

Ribosome is an essential protein-RNA complex that synthesizes proteins according to the gene library of the cell in all living organisms. In this review we will mostly focus on the

* Corresponding author e-mail: pemra@prc.boun.edu.tr.

structure-dynamics-function of the bacterial ribosome, due to the fact that most of current computational research is available for this complex.

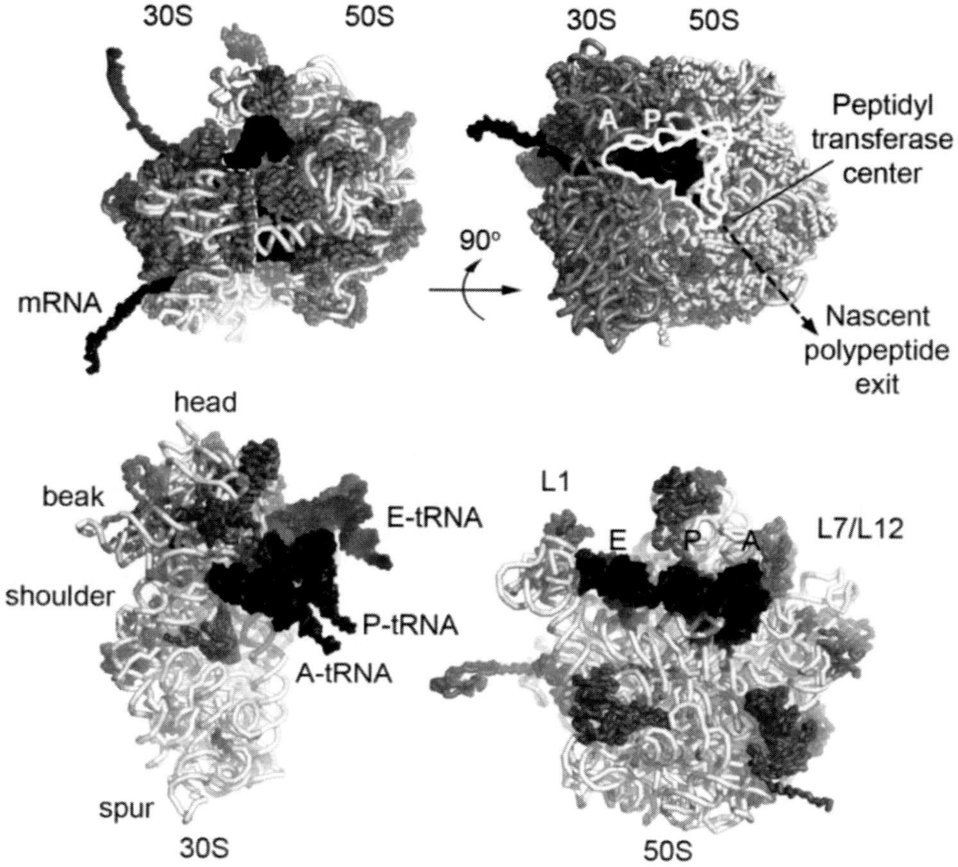

Figure 1. Bacterial ribosome complex 70S viewed from different perspectives (upper panels), and its subunits 30S and 50S viewed from the interface (lower panels). rRNAs are shown in white ribbon with r-proteins in gray ribbon. In the upper right panel, smaller subunit 30S is colored in gray to distinct it from larger subunit 50S in white. mRNA strand and A-, P, E-tRNAs are shown in black filling. Functional regions of the ribosome are indicated on the figure.

Structure of the bacterial ribosome consists of two subunits, respectively called 30S (small subunit) and 50S (large subunit), that form the 70S complex (Figure 1). Smaller subunit accommodates the 16S ribosomal RNA (rRNA, ~1550 nucleotides) and about 20 ribosomal proteins (r-proteins), and larger subunit contains the 23S rRNA (~2900 nucleotides), the 5S rRNA (~120 nucleotides) and more than 30 r-proteins.

Each ribosomal subunit carries out complementary synchronal tasks during translation of the genetic code. Smaller subunit is responsible for correct frame reading, whereas larger subunit coordinates trafficking of biomolecules coming into or out of the complex as well as catalysis of the peptide bond. During protein synthesis, messenger RNA (mRNA) strand interacts with transfer RNAs (tRNA) transporting amino acids at three specific binding sites at the interface of the subunits; A (aminoacyl), P (peptidyl) and E (exit) (Agrawal *et al* 1996) (Figure 1). Each tRNA-amino acid assisted by the prokaryotic elongation factor EF-Tu enters the complex from A-site. Anticodon of tRNA is checked for correct pairing between the mRNA codon at the decoding center of the complex on the 16S rRNA. With the recruitment of correct tRNA, ribosome catalyzes the peptide bond formation between two amino acids covalently attached to the CCA ends of A- and P-tRNAs at its peptidyl transferase center formed of rRNA on 50S. After peptide bond synthesis, the nascent polypeptide chain grows by one amino acid and proceeds through the ribosomal tunnel of ~100 Å long embedded in 50S.

Translocation of A- and P-tRNAs to their new sites P and E, respectively, is assisted by EF-G and GTP hydrolysis, while the deacylated tRNA leaves the complex from the E-site. This cycle is repeated until the mRNA stop codon is reached, resulting in the release of the nascent polypeptide chain and the separation of subunits. More details of our current understanding of bacterial ribosome and its activities during translation of genetic code into a functional protein are discussed elsewhere (Schmeing 2009, Yonath 2010, Steitz 2010, Ramakrishnan 2010).

Inasmuch as bacterial ribosome is an antibiotic target (Hermann 2005, Poehlsgaard and Douthwaite 2005), developing new effective therapeutics definitely depends on the level of our understanding of its structure and function relationship. For example, cryo-electron microscopy (cryo-EM) has provided first low-resolution pictures of the ratchet-like rotation of ribosomal subunits, which is necessary for protein synthesis (Frank and Agrawal 2000, Valle *et al* 2003, Spahn *et al* 2004). More detailed structural data, critical to fully understand the dynamic machinery and the translation process became available at the beginning of the millennium (Schluenzen *et al* 2000, Wimberly *et al* 2000, Ban *et al* 2000, Harms *et al* 2001, Yusupov *et al* 2001). Then, it became possible to investigate dynamic and functional properties of the ribosome by computational studies at various time scales and resolution levels from fine to coarse.

Full-atom molecular dynamics (MD) simulations have given insights on the functional dynamics of the ribosomal complex at atomistic detail, such as accommodation of tRNA during translation (Sanbonmatsu *et al* 2005, Withford *et al* 2010), dynamics of ribosomal A-site (Vaiana *et al* 2006, Romanowska *et al* 2008), interactions and energetics between tRNA anticodons and mRNA codons (Sanbonmatsu and Joseph 2003, Almlof *et al* 2007), catalytic mechanism of ribosome in its peptidyl transferase center (Trobro and Aqvist 2005, Sharma *et al* 2005), interdependent dynamics of L1 stalk and tRNA (Trabuco *et al* 2010a), dynamics of the ribosomal tunnel and the nascent chain (Petrone *et al* 2008, Trabuco *et al* 2010b), interactions between the complex and the membrane protein SecYE (Gumbart *et al* 2009, Frauenfeld *et al* 2011), decoding center dynamics upon drug binding (Vaiana and Sanbonmatsu 2009) and interactions between ribosome and EF-Tu (Wieden *et al* 2010, Adamczyk and Warshel 2011). A detailed up-to-date review of full-atomistic computational studies on the intact ribosome and its functional sites is compiled by Sanbonmatsu (2012).

Full-atom techniques necessitate high speed supercomputers to study functional motions of supramolecules such as the 2.5 MDa ribosome. An exemplary explicit MD study on the ribosome has simulated ~2.64 million atoms for a total sampling time of 20 ns on a supercomputer with 512 to 1024 processors (Sanbonmatsu *et al* 2005). However, computational resources may still not be sufficient to investigate such massive systems for large temporal scales.

In this respect, coarse-grained models that reduce the system size and simplify the energy function of residue interactions provide computationally efficient alternatives. Carefully interpreted, these models can reveal milliseconds to seconds level functional motions of supramolecules. In this review, we highlight important findings on the dynamics of the intact bacterial ribosome and its functional regions, elucidated by various coarse-grained approaches.

Coarse-Grained Approaches for the Intact Ribosome

Cryo-EM studies have presented first pictures of sophisticated motions for the ribosome complex (Frank and Agrawal 2000, Valle *et al* 2003, Spahn *et al* 2004). Two unequal subunits undergo a ratchet-like motion upon binding of EF-G.GTP complex, which is critical for translocation of tRNAs to their next sites together with mRNA moving exactly one codon. During the same time, coarse-grained elastic network models (ENM) based on normal mode analysis have become popular in terms of providing the functionally relevant collective motions and conformational flexibility of proteins and their complexes (Tirion 1996, Bahar *et al* 1997, Hinsen *et al* 1999, Doruker *et al* 2000, Atilgan *et al* 2001). In these simple models, the native globular protein structure is treated as the minimum energy conformation and a low-resolution elastic network is formed by connecting node pairs that fall within a predefined cutoff radius by harmonic springs. Nodes are usually placed at C^{α} atoms of amino acids and P atoms of nucleotides, therefore keeping the degrees of freedom tractable and maintaining computational efficiency for the study of very large biomolecules. At least one of the slowest normal modes obtained through vibrational analysis agrees with the functional conformational changes observed during ligand binding (Tama and Sanejouand 2001, Bahar *et al* 2007). Moreover, these global modes of motion can be obtained effectively even at extreme levels of coarse-graining, such as ten or more residues per node, as long as the three-dimensional structure of the protein is kept intact (Doruker *et al* 2002). Based on these findings and due to their computational efficiency, ENMs have been considered as a perfect choice for gaining insights about the functional dynamics of the supramolecule ribosome.

After structures became available for the subunits and the complex, ENMs have provided means for flexible fitting of atomic structures into low-resolution cryo-EM maps to obtain near-atomic conformations of the ribosome complex (Chacon *et al* 2003, Tama *et al* 2003, Tama *et al* 2004). These EM reconstructs of the ribosome further support that functionally relevant motions of supramolecules can be captured by performing vibrational analysis without the requirement of atomic structures. The ratcheting motion of the subunits along with motions of many functional regions has been analyzed in several ENM studies (Wang *et al* 2004, Kurkcuoglu *et al* 2008, Kurkcuoglu *et al* 2009a). A half millisecond long coarse-grained MD simulation (Trylska *et al* 2005a) also provides a consistent picture, where the ribosome structure is constructed using P and C^{α} atoms from the crystal structures with bonded and non-bonded interactions parameterized using harmonic and Morse potentials, respectively. Mixed levels of coarse-graining techniques could also be applied to ENMs of ribosome (Kurkcuoglu *et al* 2009a), in which functional parts such as the decoding center are described at atomistic level and higher levels of coarse-graining are used for rest of the structure to increase the computational efficiency.

Global Dynamics of the Ribosome Complex and Its Functional Parts

Computational studies have indicated that ratchet-like motion describes conformational changes at specific regions of the complex. For example, significant changes are observed at two bridges B1a and B1b at the interface of 30S and 50S, which seem to play a critical role in the control of ratcheting (Tama *et al* 2003). Important deformations are observed at the arms of tRNAs docked between 30S and 50S due to rotational motion of the subunits (Wang *et al* 2004) (Figure 2). Moreover, L1 stalk, which helps E-tRNA exit the complex, moves in phase with the rotating 30S (Trylska *et al* 2005), in accord with cryo-EM studies (Valle *et al* 2003). Interestingly, the ratcheting motion of subunits is preserved in the absence of GTP hydrolysis during MD simulations (Trylska *et al* 2005), in the absence of tRNAs (Trylska *et al* 2005, Kurkcuoglu *et al* 2008), upon removal of large stalks L1 and L7/L12 from the larger subunit or even after removal of all r-proteins with the ribosome preserving its overall shape (Kurkcuoglu *et al* 2008, Kurkcuoglu *et al* 2009a). Indeed, experimental studies have shown that protein synthesis continues in the absence of GTP hydrolysis (Gavrilova *et al* 1976, Southworth *et al* 2002) and without r-proteins (Khaitovich *et al* 1999), agreeing with computational results. Analyses of both harmonic and anharmonic motions imply that ratcheting motion of the complex is independent of atom details, partially dependent on GTP hydrolysis, and it is an intrinsic property of the shape, which coordinates functional conformational changes at various parts of ribosome.

Head rotation is reported as another significant conformational rearrangement in elastic network studies on intact ribosome (Kurkcuoglu *et al* 2008) (Figure 2). This motion has been previously monitored during tRNA selection (Ogle *et al* 2002), translation (Agrawal *et al* 1999), subunit association (Lata *et al* 1996), and tRNA positioning (Berk *et al* 2006). Head domain of the small subunit seems to be very flexible as shown in ENM studies (Tama *et al* 2003, Wang *et al* 2004, Kurkcuoglu *et al* 2008). Recently, interactions between ribosomal domains have been investigated by mapping MD simulation fluctuations onto the coarse-grained structure of ribosome, where only C^{α} and P atoms are considered (Zhang *et al* 2011). In this study, the high flexibility of the small subunit head (3' major domain in 16S rRNA) is observed to be a consequence of small number of interactions with 5', central, and 3' minor domains in 16S rRNA, whereas a large number of interactions with r-proteins imparts stability to the structure.

Head and shoulder of the smaller subunit are close to the binding site of EF-G and EF-Tu (Agrawal *et al* 1999, Agrawal *et al* 2001) located at the base of the L7/L12 stalk. This stalk is very mobile in low-frequency motions (Tama *et al* 2003, Wang *et al* 2004, Trylska *et al* 2005, Kurkcuoglu *et al* 2008), and in force-mapping studies (Zhang *et al* 2011). Moreover, the negative correlation between the large stalks L7/L12 and L1 observed in low-frequency motions imply a functional link between these distant sites (Wang *et al* 2004, Trylska *et al* 2005, Kurkcuoglu *et al* 2008). Indeed, an allosteric communication between L1 stalk and binding site of EF-G has been revealed by single molecule FRET studies during the translocation of tRNAs (Feia *et al* 2009). Focusing on L1 dynamics indicates that L1 stalk is also very mobile and it opens and closes a gate where the E-tRNA can leave the complex (Trylska *et al* 2005), and also affects E-tRNA dynamics (Kurkcuoglu *et al* 2008). It is now widely accepted that two stalks on the larger subunit coordinate the recruitment and release of biomolecules into and out of the complex.

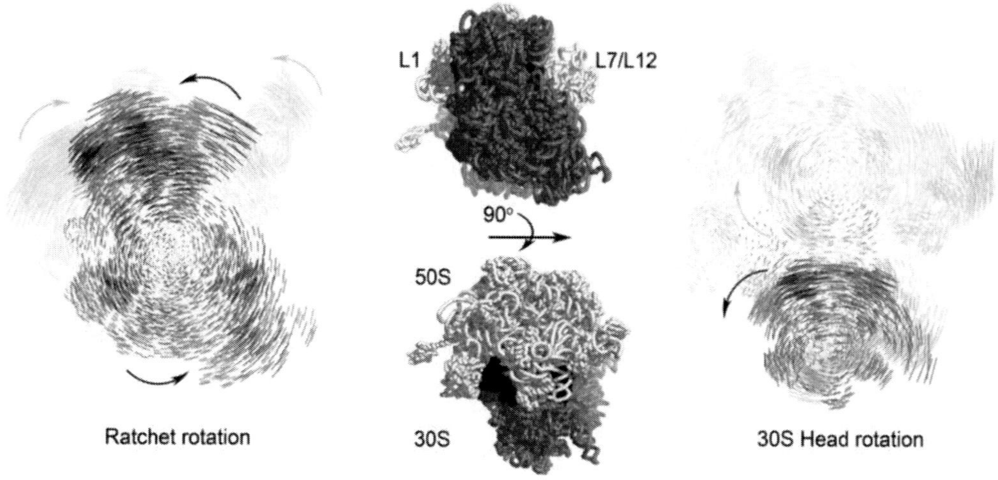

Figure 2. Ratchet rotation and small subunit head rotation calculated with ENM using normal mode analysis (Kurkcuoglu *et al* 2008) are shown with displacement vectors of residues. Black and gray arrows indicate one alternative direction of the harmonic motion of small and large subunits, respectively.

Calculations of changes in electrostatic properties of the translating ribosome have provided useful insights on protein synthesis mechanism (Trylska *et al* 2005b). Normal modes of the complex using only P and C^{α} atoms have been employed together with a reduced Poisson-Boltzmann potential energy in order to calculate the electrostatic potential of the ratcheting structure. The charge distribution on the complex appears to be mainly electronegative due to large amount of rRNA forming the machinery scaffold, agreeing with a previous all-atom study (Baker *et al* 2001). Positive patches are located especially at the binding sites of elongation factors, release factors and ribosomal protein L1 interacting with electronegative E-tRNA. Ratcheting motion of the complex is observed to change electrostatic properties of tRNA binding sites, underlying the importance of long-range interactions in binding processes and in translation.

Motions of tRNAs and Translocation

During protein synthesis, tRNAs translocate on the large complex along with mRNA, in order to vacate the A-site for a newly recruited tRNA-amino acid pair. Promoted by EF-G and GTP hydrolysis, and assisted with the ratchet motion of subunits, tRNAs move from one site to another, while visiting hybrid states such as P/E, in which the tRNA is partially bound to the P-site on smaller subunit and to the E-site on larger subunit. During their accommodation on the ribosome, tRNAs change their conformation and build new interactions with the complex, which have been monitored by extensive experimental and explicit MD studies (Sanbonmatsu *et al* 2005, Li and Frank 2007, Berk and Cate 2007, Julian *et al* 2008, Munro *et al* 2009, Whitford *et al* 2010).

Coarse-grained studies on intact ribosome have focused on internal deformations on tRNA structures and on their low-frequency motions related with translocation, in order to understand the translocation mechanism. A-tRNA and E-tRNA have been observed to undergo significant internal deformations especially at their arm regions during slowest motions of ribosome, whereas mean-square fluctuations of P-tRNA, as well as peptidyl and aminoacyl ends of tRNAs, are quite small (Wang *et al* 2004, Wang and Jernigan 2005, Kurkcuoglu *et al* 2008). This indicates that the ribosome structure geometrically confines tRNAs at the peptidyl transferase center and secures the peptide synthesis to grow the nascent polypeptide at a certain rate. On the other hand, A-tRNA and P-tRNA exert correlated motions, while E-tRNA moves along with L1 stalk (Wang *et al* 2004, Wang and Jernigan 2005, Kurkcuoglu *et al* 2008). This observation suggests that low-frequency motions of the complex, defined by the ribosomal architecture, are critical in directing A- and P-tRNAs to their next sites and in releasing deacylated E-tRNA.

Helicase Activity of the Ribosome

Single stranded mRNA is highly flexible and can easily adopt secondary structures before entering the ribosome complex. An earlier study has revealed that r-proteins S3 and S4, which form a tight entrance together with S5 on smaller subunit, carry a helicase activity for the ribosome by disrupting the downstream helices into a straight mRNA conformation for correct frame-reading (Takyar *et al* 2005). Mutations on Arg and Lys residues on r-proteins S3 and S4 close to mRNA strand impair the helicase activity of ribosome indicating the importance of electrostatic interactions for mRNA unwinding. In addition, ribosome helicase can still function in the absence of GTP hydrolysis and elongation factors, pointing that the energy source for disrupting mRNA strand comes from conformational changes of the complex during translocation. Further mutations on ribosome reveal that S5 also plays an important role in translation fidelity by orienting mRNA (Kirthi *et al* 2006).

Later, ENM has suggested that several polar and hydrophilic residues on S3 and S4 encircling the tight entrance of the mRNA to the complex may also bear a role in the helicase activity (Kurkcuoglu *et al* 2008). Although these residues are close to the highly mobile regions of 30S, such as the head and beak, they stay at a stable geometry with significantly low mean-square fluctuations in the functionally important low-frequency modes. In addition, binding of EF-Tu to the translating complex seems to further stabilize fluctuations of residues on proteins S3, S4 and S5 staying close to mRNA, and to control mRNA thermal fluctuations for correct frame-reading. Similarly, exit region of the mRNA from the complex, which is encircled with r-proteins S7, S11 and anti Shine-Dalgarno strand on 16S rRNA, exhibits low amplitude fluctuations at slow modes. These findings (Kurkcuoglu *et al* 2008) have indicated a symmetrical clamp to the mRNA entrance, which may act as a reel to help the translocation of mRNA during ratcheting motion. A recent experimental study has confirmed these observations and revealed that ribosome helicase biases the thermal fluctuations of the single strand at an open state, and mechanically pulls the mRNA apart to unwind it by large conformational changes during translocation, while ensuring a basal translation rate (Qu *et al* 2011).

Dynamics of Ribosomal Tunnel and Nascent Polypeptide

The ribosomal tunnel is mainly formed of 23S rRNA with some protrusions of r-proteins (Figure 3). Embedded in the larger ribosomal subunit, the tunnel is ~100 Å long with a varying diameter of ~10-20 Å. The narrowest area on the tunnel close to the peptidyl transferase center functions as a discriminating gate directed by the highly conserved r-proteins L4 and L22 to stall translation process for specific peptide chains (Nakatogawa and Ito 2002, Tenson and Ehrenberg 2002).

Confinement and crowding are two important parameters that increase the binding rate of molecules to other molecules and surfaces, as well as the folding rate of polypeptide chains (Zhou *et al* 2008). Along the same line, a Langevin dynamics study has revealed that the nascent chain folding inside the ribosomal tunnel is entropically driven (Ziv *et al* 2005). Here, the tunnel is simply modeled as a confined cylinder, and the polypeptide as a chain of hydrophilic or hydrophobic beads placed at the C^{α} positions of residues. The confinement effect on folding has been proposed to maximize through a passage wide enough to accommodate an α-helical structure, and to stabilize longer helix-forming chains than shorter ones. Under a certain diameter, a folded structure into a β-hairpin can be destabilized and adopt an extended conformation, as suggested by a similar Langevin dynamics study (Kirmizialtin *et al* 2004).

Similarly, a kinetic Monte Carlo simulation of an on-lattice model (Contreras Martinez *et al* 2006) has monitored folding behaviors of various polypeptide chains in confined rectangular tunnels of different sizes and chemical composition (hydrophobic, hydrophilic or inert walls). Simulations indicate that the folding of a so-called 'well-designed' protein may be facilitated by confinement depending on the specific parameters employed. Considering a varying diameter of the exit tunnel between 10 and 20 Å, folding of a nascent chain into a secondary structure can be effected by its location through the tunnel. However, a secondary structure with β-sheets or a tertiary structure would not seem to fit due to geometry of the tunnel (Voss *et al* 2006). Brownian dynamics of co-translational folding, i.e. folding of the nascent chain while it is being synthesized, has been studied for three small proteins within the large ribosomal subunit using one bead Go-like model (Go 1983, Elcock 2006). Here, the flexible nascent chain is described at residue-level by its C^{α} atoms and the rigid large subunit is described at near-atomic level.

Beads are added at specific rates to the C-terminus of the chains at the peptidyl transferase center for modeling protein synthesis. Analysis of chain conformations has indicated that confinement does not seem to have a significant effect on folding, at least for those chains adopting dominantly β-sheet structures. On the other hand, if there is some degree of stable structure formation within the tunnel, this may delay the rapid release of protein from the complex.

The effect of macromolecular crowding on co-translational folding in the absence of a chaperone has been studied for small domains using a coarse-grained Langevin dynamics simulation (Jefferys *et al* 2010), where each amino acid is represented by two nodes located at C^{α} atom and side chain centroid. The chain emerging from the tunnel that is released to a crowded environment in simulations is capable of folding into native-like structures up to levels of 40% excluded volume. Beyond this limit, which is also consistent that observed in the cell, the chains lose their conformational freedom and thus cannot attain native-like structures.

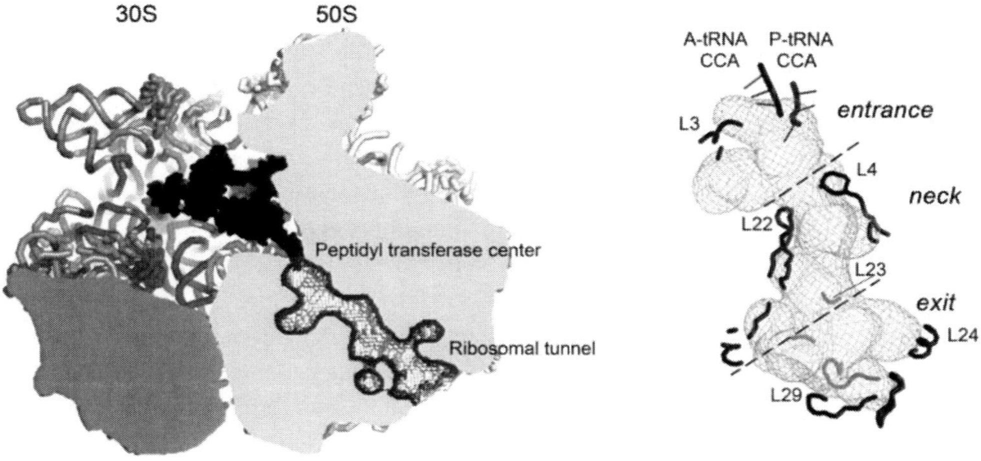

Figure 3. Ribosomal tunnel cavity extracted from the 5.5 Å resolution ribosome crystal structure (Yusupov *et al* 2001) using the software ^{3}V (Voss *et al* 2006) is shown on the complex on the left. On the right, the tunnel is shown in gray mesh surface, and nearby residues of r-proteins within 10 Å and CCA ends of A- and P-tRNAs are colored in black. 23S rRNA is not shown for clarity. Three dynamically distinct regions (entrance, neck, exit) determined with ENM (Kurkcuoglu *et al* 2009b) are indicated on the figure.

The dynamics of the ribosomal tunnel has also been studied via coarse-grained ENM of the intact residue-level ribosome structure (Kurkcuoglu *et al* 2009b). All possible combinations of the five lowest-frequency modes are used in order to visualize dynamics of residues forming the tunnel. As the tunnel is deeply embedded in the larger subunit, mixed modes exhibit quite complex dynamics that is distinct from other structures with cavities, such as GroEl-GroES with shearing and stretching motions. Correlation analysis reveals three distinct zones along the tunnel; an entrance region moving dominantly in the nascent chain exit direction, a neck region exerting rotational motion, and an exit region fluctuating with mixed rotational and perpendicular motions to lead the nascent chain to the exit (Figure 3). These three dynamic regions are similar to experimentally determined three folding zones along the tunnel (Lu and Deutsch 2005). Interestingly, regions close to the peptidyl transferase center and to the exit have higher impact on polypeptide folding as compared to the middle zone. In addition, slow modes and their linear combinations indicate that at the narrow gate, proteins L4 and L22 display anti- or non-correlated motions, which may have a role in the gating mechanism.

Ribosome Assembly

The accurate assembly of the ribosome subunits into a functional ribosome structure is critical for synthesizing new proteins and cell growth. A detailed assembly map providing the sequence of binding of r-proteins to the smaller ribosomal subunit has been determined *in vitro* almost four decades ago for *Escherichia coli* (Held *et al* 1974). However, our current level of understanding on assembly mechanism remains limited. Thus, coarse-grained techniques have been employed to investigate the assembly process by focusing on structural rearrangements in the ribosome subunits, especially on 30S due to rich experimental data.

The contribution of the binding of r-proteins to the 30S folding has been studied using coarse-grained Monte Carlo simulated annealing technique (Stagg *et al* 2003). Interaction between two pseudo-atoms representing either an amino acid or a nucleotide, are modeled by bonded harmonic and non-bonded semi-harmonic volume exclusion terms, respectively. Protein-DNA contacts are determined from crystal structure and used as a restraint during rigid body simulations. Simulations show that binding of early-binding proteins to the rRNA has an impact on the pre-organization of the remaining unoccupied sites.

Relative binding free energies of 30S proteins to the 16S rRNA have been calculated based on a Poisson-Boltzman implicit solvent model (Trylska *et al* 2005b). Although the model has limitations such as using a rigid body model, and poor quality of the crystal structure that can affect the calculations, it provides useful insights on the role of charge-charge interactions during the assembly process. Calculations show that the 5' domain early kinetic class r-proteins display most favorable binding and late-binding proteins do not prefer to bind to naked 16S rRNA, consistent with earlier experimental observations.

Changes in conformational dynamics of the small ribosomal subunit due to removal of r-proteins either one at a time, or groups have been investigated by ENM (Yan *et al* 2008). Coarse-grained normal mode analysis indicates that r-proteins S6 and S18 have interdependence and display strong positive correlation of their motions. These dynamically correlated proteins form a stable heterodimer before assembly (Recht and Williamson 2001) and calculations suggest that they also function as a block in the 30S structure.

The same group has recently studied the dynamics properties of the 30S proteins in order to find a connection between their conformational changes and binding to 16S rRNA during assembly (Burton *et al* 2012). Interestingly, rRNA contact residues on r-proteins are found to be highly conserved compared to the sequence of full-length protein, and to mainly consist of charged and polar amino acids that can make hydrogen bond with rRNA. Both atomistic MD and coarse-grained ENM consistently show that most r-proteins contain mobile residues enriched at the rRNA contact regions. The degree of mobility depends on the structure of the r-protein; proteins exposed to solvent in the 30S subunit adopt more compact conformation in solution, whereas proteins anchored to the 16S rRNA by long extended and embedded domains display more flexibility.

Conclusion

Coarse-graining biomolecules to study their functional dynamics remains an attractive approach since system sizes of interest increase with our curiosity for understanding cellular activities, therefore may stay beyond the capability of conventional full-atom techniques due to limitations in computational power. In this perspective, coarse-grained techniques have provided significant findings and useful predictions on the functional mechanisms employed by the gigantic molecular machine ribosome, thanks to the rich accumulating structural data. Details of the ribosome structure and energy functions to describe residue interactions may differ from a model to another; but nearly all coarse-graining approaches recognize the importance of studying ribosome as a whole -considered as either rigid or flexible- to reveal its complicated functional motions and mechanisms used for protein synthesis.

All findings and calculations on ribosome complex reviewed here and elsewhere (Trylska 2010, Sanbonmatsu 2012) expose the relationship between the sophisticated structure and complicated dynamics of the molecular machine. Ribosome's fascinating architecture guaranties an open conformation of mRNA before it is decoded, a geometrically confined region for correct frame-reading, a highly conserved rigid rRNA cavity for peptide bond synthesis, a translocation mechanism to advance tRNAs exactly one site and the mRNA one codon, and a long dynamic exit tunnel to partially fold and guide the nascent chain, altogether orchestrated at the same time or sequentially on relatively close regions of the complex. Moreoever, electrostatic properties of the ribosome structure are carefully designed for effective binding of biomolecules to the complex from specific spots, which is important for the ribosome to follow a basal rate for protein synthesis with least error.

There are surely many intriguing questions waiting to be answered about ribosome machinery. One of them is understanding how site-to-site communication through ribosome structure is maintained during protein synthesis. For example, a recent mutation study investigated the role of the B1 b/c inter-subunit bridge of yeast ribosome in translational fidelity (Rhodin and Dinman 2011). Observations suggested that r-protein S18, L11 side of bridge B1 b/c and helix H84 lie on a communication network between the decoding center and peptidyl transferase center of the complex. Many other pathways on ribosome can be explored with coarse-grained models focusing on allostery, which can further propose new potential drug binding sites to inhibit bacterial growth.

Acknowledgments

PD acknowledges partial support by DPT project (no. 2009k120520) and the Betil Fund.

References

Adamczyk, A. J.; Warshel, A. *Proc. Natl. Acad. Sci. USA*. 2011, 108, 9827-9832.

Agrawal, R. K.; Penczek, P.; Grassucci, R. A.; Li, Y.; Leith, A.; Nierhaus, K. H.; Frank, J. *Science*. 1996, 271, 1000-2.

Agrawal, R. K.; Heagle, A. B.; Penczek, P.; Grassucci, R. A.; Frank, J. *Nat. Struct. Biol.* 1999, 6, 643–647.

Agrawal, R. K.; Linde, J.; Sengupta, J.; Nierhaus, K. H.; Frank, J. *J. Mol. Biol.* 2001, 311, 777–787.

Almlof, M.; Ander, M.; Aqvist, J. *Biochemistry*. 2007, 46, 200-209.

Atilgan, A. R.; Durell, S. R.; Jernigan, R. L.; Demirel, M. C.; Keskin, O.; Bahar I. *Biophys. J.* 2001, 80:505–515.

Bahar, I.; Atilgan, A. R.; Erman, B. *Fold. Des.* 1997, 2, 173–181.

Bahar, I.; Chennubhotla, C.; Tobi, D. *Curr. Opin. Struct. Biol.* 2007, 17, 633-640.

Baker, N. A.; Sept, D.; Joseph, S.; Holst, M. J.; MCCammon, J. A. *Proc. Natl. Acad. Sci. USA*. 2001, 98, 10037–10041.

Ban, N.; Nissen, P.; Hansen, J.; Moore, P. B.; Steitz, T. A. *Science.* 2000, 289, 905–920.

Berk, V.; Zhang, W.; Pai, R. D.; Cate, J. H. *Proc. Natl. Acad. Sci. USA.* 2006, 103, 15830-15834.

Berk, V.; Cate, J. H. D. *Curr. Opin. Struct. Biol.* 2007, 17, 302–309.

Burton, B.; Zimmermann, M. T.; Jernigan, R. L.; Wang, Y. *PLoS Comput. Biol.* 2012, 8(5), e1002530.

Chacon, P.; Tama, F.; Wriggers, W. *J. Mol. Biol.* 2003, 326, 485–492.

Contreras Martinez, L. M.; Martinez-Veracoechea, F. J.; Pohkarel, P.; Stroock, A. D.; Escobedo, F. A.; DeLisa, M. P. *Biotechnol. Bioeng.* 2006, 94(1), 105-117.

Doruker, P.; Atilgan, A. R.; Bahar, I. *Proteins.* 2000, 40, 512-524.

Doruker, P.; Jernigan, R. L.; Bahar, I. *J. Comput. Chem.* 2002, 23:119–127.

Elcock, A. H. *PloS Comput. Biol.* 2006, 2(7), e98.

Feia, J.; Bronsona, J. E.; Hofmanb, J. M.; Srinivasc, R. L.; Wigginsd, C. H.; Gonzalez, R. L. Jr. *Proc. Natl. Acad. Sci. USA.* 2009, 106(37), 15702-15707.

Frank, J.; Agrawal, R. K. *Nature.* 2000, 406, 318-322.

Frauenfeld, J.; Gumbart, J.; van der Sluis, E. O.; Funes, S.; Gartmann, M.; Beatrix, B.; Mielke, T.; Berninghausen, O.; Becker, T.; Schulten, K.; Beckmann, R. *Nat. Struct. Mol. Biol. 2011*, 18, 614-621.

Gavrilova, L. P.; Kostiashkina, O. E.; Koteliansky,V. E.; Rutkevitch, N. M.; Spirin, A. S. *J. Mol. Biol.* 1976, 101, 537–552.

Go, N. *Annu. Rev. Biophys. Bioeng.* 1983, 12, 183–210.

Gumbart, J.; Trabuco, L. G.; Schreiner, E.; Villa, E.; Schulten, K. *Structure. 2009*, 17, 1453-1464.

Harms, J.; Schluenzen, F.; Zarivach, R.; Bashan, A.; Gat, S.; Agmon, I.; Bartels, H.; Franceschi, F.; Yonath, A. *Cell.* 2001, 107, 679–688.

Held, W. A.; Ballou, B.; Mizushima, S.; Nomura, M. *J. Biol. Chem.* 1974, 249(10), 3103-3111.

Hermann, T. *Curr. Opin. Struct. Biol.* 2005, 15, 355–366.

Hinsen, K.; Thomas, A.; Field, M. J. *Proteins.* 1999, 34, 369-382.

Jefferys, B. R.; Kelley, L. A.; Sternberg, M. J. E. *J. Mol. Biol.* 2010, 397, 1329-1338.

Julián, P.; Konevega, A. L.; Scheres, S. H. W.; Lázaro, M.; Gil, D.; Wintermeyer, W.; Rodnina, M. V.;Valle, M. *Proc. Natl. Acad. Sci. USA.* 2008, 105(44), 16924-16927.

Khaitovich, P.; Mankin, A. S.; Green, R.; Lancester, L.; Noller, H. F. *Proc. Natl. Acad. Sci. USA.* 1999, 96, 85–90.

Kirmizialtin, S.; Ganesan, V.; Makarov, D. E. *J. Chem. Phys.* 2004, 121, 10268–10277.

Kirthi, N.; Roy-Chaudhuri, B.; Kelley, T.; Culver, G. M. *RNA.* 2006, 12, 1–12.

Kurkcuoglu, O.; Doruker, P.; Sen, T. Z.; Kloczkowski, A.; Jernigan, R. L. *Phys. Biol.* 2008, 5, 046005.

Kurkcuoglu, O.; Turgut, O. T.; Cansu, S.; Jernigan, R. L.; Doruker, P. *Biophys. J.* 2009a, 97, 1178–1187.

Kurkcuoglu, O.; Kurkcuoglu, Z.; Doruker, P.; Jernigan, R. L. *Proteins.* 2009b, 75, 837–845.

Lata, K. R.; Agrawal, R. K.; Penczek, P.; Grassucci, R.; Zhu, J.; Frank, J. *J. Mol. Biol.* 1996, 262, 43–52.

Li, W.; Frank, J. *Proc. Natl. Acad. Sci. USA.* 2007, 104(42), 16540-16545.

Lu, J.; Deutsch, C. *Nat. Struct. Mol. Biol.* 2005, 12,1123–1129.

Munro, J. B.; Sanbonmatsu, K. Y.; Spahn, C. M. T.; Blanchard, S. C. *Trends Biochem. Sci.* 2009, 34(8), 390-400.

Nakatogawa, H.; Ito, K. *Cell.* 2002, 108, 629–636.

Ogle, J. M.; Murphy, F. V.; Tarry, M. J.; Ramakrishnan, V. *Cell.* 2002, 111, 721–732.

Petrone, P. M.; Snow, C. D.; Lucent, D.; Pande, V. S. *Proc. Natl. Acad. Sci. USA.* 2008, 105(43), 16549-16554.

Poehlsgaard, J.; Douthwaite, S. *Nat. Rev. Microbiol.* 2005, 3, 870–881.

Qu, X.; Wen, J.-D.; Lancaster, L.; Noller, H. F.; Bustamante, C.; Tinoco, I. Jr. *Nature.* 2011, 475, 118-121.

Ramakrishnan, V. *Angew. Chem. Int. Edn.* 2010, 49(26), 4355-4380.

Recht, M. I.; Williamson, J. R. *J. Mol. Biol.* 2001, 313, 35–48.

Romanowska, J.; Setny, P.; Trylska, J. *J. Phys. Chem. B.* 2008, 112, 15227–15243.

Sanbonmatsu, K. Y.; Joseph, S. *J. Mol. Biol.* 2003, 328, 33-47.

Sanbonmatsu, K. Y.; Joseph, S.; Tung, C. S. *Proc. Natl. Acad. Sci. USA.* 2005, 102, 15854-15859.

Sanbonmatsu, K. Y. *Curr. Opin. Struct. Biol.* 2012, 22, 168–174.

Schluenzen, F.; Tocilj, A.; Zarivach, R.; Harms, J.; Gluehmann, M.; Janell, D.; Bashan, A.; Bartels, H.; Agmon, I.; Franceschi; Yonath, A. *Cell.* 2000, 102, 615–623.

Schmeing, T. M.; Ramakrishnan, V. *Nature.* 2009, 461, 1234-1242.

Sharma, P. K.; Xiang, Y.; Kato, M.; Warshel, A. *Biochemistry.* 2005, 44, 11307-11314.

Southworth, D. R.; Brunelle, J. L.; Green, R. *J. Mol. Biol.* 2002, 324, 611–623.

Spahn, C. M. T.; Gomez-Lorenzo, M. G.; Grassucci, R. A.; Jorgensen, R.; Andersen, G. R.; Beckmann, R.; Penczek, P. A.; Ballesta, J. P.; Frank, J. *EMBO J.* 2004, 23, 1008–1019.

Stagg, S. M.; Mears, J. A.; Harvey, S. C. *J. Mol. Biol.* 2003, 328, 49–61.

Steitz, T. A. *Angew. Chem. Int. Edn.* 2010, 49(26), 4381-4398.

Takyar, S.; Hickerson, R. P.; Noller, H. F. *Cell.* 2005, 120, 49–58.

Tama, F.; Sanejouand, Y. H. *Protein Eng.* 2001, 14(1), 1-6.

Tama, F.; Valle, M.; Frank, J.; Brooks, C. L. III. *Proc. Natl. Acad. Sci. USA.* 2003, 100(16), 9319–9323.

Tama, F.; Miyashita, O.; Brooks, C. L. III. *J. Struct. Biol.* 2004, 147, 315–26.

Tenson, T.; Ehrenberg, M. *Cell.* 2002, 108, 591–594.

Tirion, M. M. *Phys. Rev. Lett.* 1996, 77, 1905–1908.

Trabuco, L. G.; Schreiner, E.; Eargle, J.; Cornish, P.; Ha, T.; Luthey- Schulten, Z.; Schulten, K. *J. Mol. Biol.* 2010a, 402, 741-760.

Trabuco, L. G.; Harrison, C. B.; Schreiner, E.; Schulten, K. *Structure.* 2010b, 18, 627-637.

Trobro, S.; Aqvist, J. *Proc. Natl. Acad. Sci. USA.* 2005, 102, 12395-12400.

Trylska, J.; Tozzini, V.; McCammon, J. A. *Biophys. J.* 2005a, 89, 1455–1463.

Trylska, J.; McCammon, J. A.; Brooks, C. L. III. *J. Am. Chem. Soc.* 2005b, 127, 11125-11133.

Vaiana, A. C.; Westhof, E.; Auffinger, P. *Biochimie.* 2006, 88, 1061-1073.

Vaiana, A. C.; Sanbonmatsu, K. Y. *J. Mol. Biol.* 2009, 386, 648-661.

Valle, M.; Zavialov, A.; Sengupta, J.; Rawat, U.; Ehrenberg, M.; Frank, J. *Cell.* 2003, 114, 123-134.

Voss, N. R., Gerstein, M.; Steitz, T. A.; Moore, P. B. *J. Mol. Biol.* 2006, 360, 893–906.

Wang, Y.; Rader, A. J.; Bahar, I.; Jernigan, R. L. *J. Struct. Biol.* 2004, 147, 302–314.

Wang, Y.; Jernigan, R. L. *Biophys. J.* 2005, 89, 3399–3409.

Whitford, P. C.; Geggier, P.; Altman, R. M.; Blanchard, S. C.; Onuchic, J. N.; Sanbonmatsu, K. Y. *RNA*. 2010, 16, 1196-1204.

Wieden, H. J.; Mercier, E.; Gray, J.; Steed, B.; Yawney, D. *Biophys. J.* 2010, 99, 3735-3743.

Wimberly, B. T.; Brodersen, D. E.; Clemons, W. M.; Morgan-Warren, R. J.; Carter, A. P.; Vonrhein, C.; Hartsch, T.; Ramakrishnan, V. *Nature*. 2000, 407, 327–339.

Yan, A.; Wang, Y.; Kloczkowski, A.; Jernigan, R. L. *J. Chem. Theory Comput.* 2008, 4(10), 1757–1767.

Yonath, A. *Angew. Chem. Int. Edn.* 2010, 49(26), 4340–4354.

Yusupov, M. M.; Yusupova, G. Z.; Baucom, A.; Lieberman, K.; Earnest, T. N.; Cate, J. H.; Noller, H. F. *Science*. 2001, 292, 883–896.

Zhang, Z.; Sanbonmatsu, K. Y.; Voth, G. A. *J. Am. Chem. Soc.* 2011, 133, 16828–16838.

Zhou, H.-X.; Rivas, G.; Minton, A. P. *Annu. Rev. Biophys.* 2008, 37, 375–397.

Ziv, G.; Haran, G.; Thirumalai, D. *Proc. Natl. Acad. Sci. USA*. 2005, 102, 18956–18961.

In: Ribosomes
Editors: Zhou Lin and Wang Liu

ISBN: 978-1-62417-698-2

Chapter 8

Ribosomal Incorporation of Unusual Amino Acids: Learning from Mother Nature

Wei Ren and Hui-wang Ai*
Department of Chemistry, University of California, Riverside, CA, US

Abstract

In living cells and organisms, proteins are often made through decoding genetic information stored as sequences of nucleotides. The central machinery in the process is ribosome. Every protein is translated from the 5' end to the 3' end of an mRNA. Starting from the translation initiation site, every three consecutive nucleotides are considered as a codon and read as a single amino acid. Cellular proteins, which are made from 61 sense codons of a total 64 codons, typically comprise 20 natural amino acids. However, there are unusual cases existing in nature. Under certain circumstances, nature can use codons that are typically nonsense to code natural as well as unusual amino acids (*e.g.* selenocysteine and pyrrolysine). In addition, ribosomal frameshifting is also not uncommon, in which case one or two nucleotides in a coding sequence are skipped to make proteins based on a shifted frame. In the past decade or so, researchers have invented procedures in laboratories to biosynthesize proteins containing unusual amino acids. Such exploration has made a big impact. This chapter summarizes methods used in nature and in laboratories to incorporate unusual amino acids into proteins. In particular, recent studies, which re-engineered ribosome and other related translational factors, are emphasized. Therapeutic applications of this technology are also discussed.

1. Ribosomal Protein Synthesis

Genetic information is mainly stored in cells as sequences of nucleotides.(1) Each nucleotide is composed of a pentose (5-carbon carbohydrate), a phosphate group extending

* To whom correspondence should be addressed. Email: huiwang.ai@ucr.edu. Fax: 1-951-750-5138.

from 5' (or 3') position of the pentose, and one of four types of nucleobases. In deoxyribonucleotides (DNA), 2-deoxyribose is the pentose, and adenine (A), guanine (G), thymine (T) and cytosine (C) are the four types of bases. Prokaryotic and eukaryotic cells use regions of DNA sequences as the templates to synthesize strands of ribonucleotides (RNA). Sequences of RNA strands are copied from DNA strands, except that ribose replaces 2-deoxyribose and uracil (U) replaces thymine as one of the four RNA bases. This process is termed as "transcription [1]". Next, Proteins are synthesized from transcribed messenger RNAs (mRNA): every three bases in an mRNA open reading frame are "translated" into a single amino acid residue. An important group of enzymes, aminoacyl transfer RNA (tRNA) synthetases, catalyze the linkage between amino acids and tRNAs. Every tRNA has a 3-base anticodon in its anticodon loop to pair with mRNA during ribosomal protein synthesis.

Ribosomes are large RNA and protein containing machineries (up to several million Da), catalyzing the formation of peptides from individual amino acids [2]. Ribosomes exist in all archaeal, eubacterial and eukaryotic cells. Although differing in size and in detailed composition, each ribosome has two subunits, a large subunit catalyzing peptidyl transfer reaction and a small subunit critical for translation initiation [2]. Translation initiation factors assemble the small subunit and the mRNA to start the formation of a translation complex. The Shine-Dalgarno (SD) sequence of prokaryotic mRNA and 5' cap of eukaryotic mRNA are very important for the initiation [3, 4]. The nearby AUG codon is then identified by the ribosome and decoded as an N-terminal *N*-formylmethionine (fMet) in prokaryotes or methionine (Met) in eukaryotes. After the ribosome is fully assembled at the initiation AUG site, it contains three RNA-binding sites, designated A, P and E sites. Elongation starts when the fMet-tRNA (or Met-tRNA in eukaryotes) enters the P site, resulting in a conformational change which opens the A site for another aminoacyl-tRNA to enter. Peptide formation is catalyzed by the ribosomal RNA in the large subunit. After the bond is formed, the A site contains a newly formed peptide, while the P site contains an uncharged tRNA. The ribosome moves along the mRNA, so the uncharged tRNA enters the E site and then exits from the ribosome. The peptidyl-tRNA enters the P site and opens the A site for the next round of coupling. Elongation factors are needed in this process, for example, to facilitate the entry of aminoacyl tRNA into the A site. When the ribosome reaches one of the three termination codons (UAA, UAG and UGA), releasing factors (proteins) would enter the A site and trigger the hydrolysis of the ester bond in peptidyl-tRNA at the P site [5]. After releasing the peptide, the whole complex is disassembled with the aid of several protein factors to recycle translation components. More detailed process about ribosomal protein synthesis can be found in recently published review articles and in other chapters of this book [6-8].

Under most circumstance, every three consecutive bases following the starting AUG codon are translated into one amino acid. All four types of nucleobases can make 64 codons. With three exceptions (UAA, UAG and UGA as stop codons), each codon encodes one of the 20 common natural amino acids. So there are degenerated codons: most of the 20 amino acids are encoded by more than one codons. The correspondence between codons, and amino acids and translational termination message, is nearly universal among all domains of life [9].

We here discuss a few exceptions. Mitochondrial ribosomes synthesize mitochondrial proteins based on different codon tables [10]. Mitochondria carry their own genome, which includes mitochondrial tRNAs. The mitochondrial genetic code has drifted from the universal code. Furthermore, organisms including bacteria, yeast and other eukaryotes can harbor suppressor tRNAs that can recognize and decode nonsense codons (UAA, UAG or UGA)

[11]. These tRNAs were most likely derived from normal tRNAs through anticodon mutations. New codon-anticodon interactions are established to read through stop codons. In most natural cases, one of the 20 common natural amino acids is inserted in response to stop codons.

It is quite unique to insert the unusual amino acid, pyrrolysine, in response to UAG codons [12, 13]. Certain methanogenic archaea including *Methanosarcina barkeri* and *M. mazei*, and the gram positive bacterium *Desulfitobacterium hafniense*, express amber suppressor tRNAs ($tRNA^{Pyl}$) and synthetases that catalyze the charge of tRNAs with pyrrolysine (Figure 1A). They also harbor gene clusters to biochemically synthesize the amino acid pyrrolysine [14, 15]. The process to insert pyrrolysine is similar to the process for ribosomal insertion of other amino acids: pyrrolysine-charged tRNAs are brought into ribosomes by typical elongation factors to extend the nascent peptides.

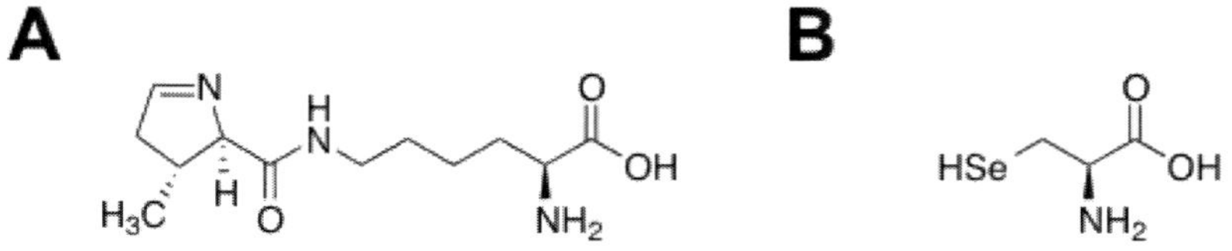

Figure 1. Chemical structures of (A) pyrrolysine and (B) selenocysteine.

Another unusual amino acid, selenocysteine (Figure 1B), is also genetically encoded in many natural organisms [16, 17]. Compared to cysteine, selenocysteine has a lower pK_a and a higher reduction potential, so diselenium bonds are more easily formed [18]. Selenocysteine has been found to play a critical role for the function of a few anti-oxidant proteins. Unlike other 20 natural amino acids and pyrrolysine, selenocysteine is not directly charged to its tRNA (Figure 2A), because there is no free selenocysteine in cells [19]. Instead, seryl-tRNA synthetase first links serine to a special selenocysteine tRNAs ($tRNA^{Sec}$). The resulting Ser-$tRNA^{Sec}$ is not recognized by translation factors, so are not used for ribosomal translation. Next, the tRNA-bound seryl residue is converted to a selenocysteine in the presence of appropriate enzymes and selenium donor molecules [19]. Alternative translational elongation factors are needed to bring selenocysteine-charged $tRNA^{Sec}$ (Sec-$tRNA^{Sec}$) into ribosome for protein synthesis (Figure 2B) [17]. The anticodon of $tRNA^{Sec}$ is UCA, so it can pair with the UGA opal codon. Not all UGA codons are suppressed, however. The mRNAs of selenocysteine-containing proteins (selenoproteins) often contain sequences called SECIS (selenocysteine insertion sequence) elements. The SECIS elements are defined by characteristic nucleotide sequences, secondary structures and base-pairing patterns. In bacteria, SECIS elements are typically located immediately after UGA codons in reading frames. In archaea and eukaryotes, SECIS elements are in the 3'-UTRs (untranslated regions) of mRNAs, and can direct multiple selenocysteines into a single peptide in response to multiple UGA codons (Figure 2B) [20]. Sec-$tRNA^{Sec}$ specific elongation factors can bind SECIS elements, and promote the delivery of Sec-$tRNA^{Sec}$ into ribosomes associated with the same mRNA. When cells are grown in the presence of selenium, corresponding UGA codons are suppressed to synthesize full-length functional selenoproteins.

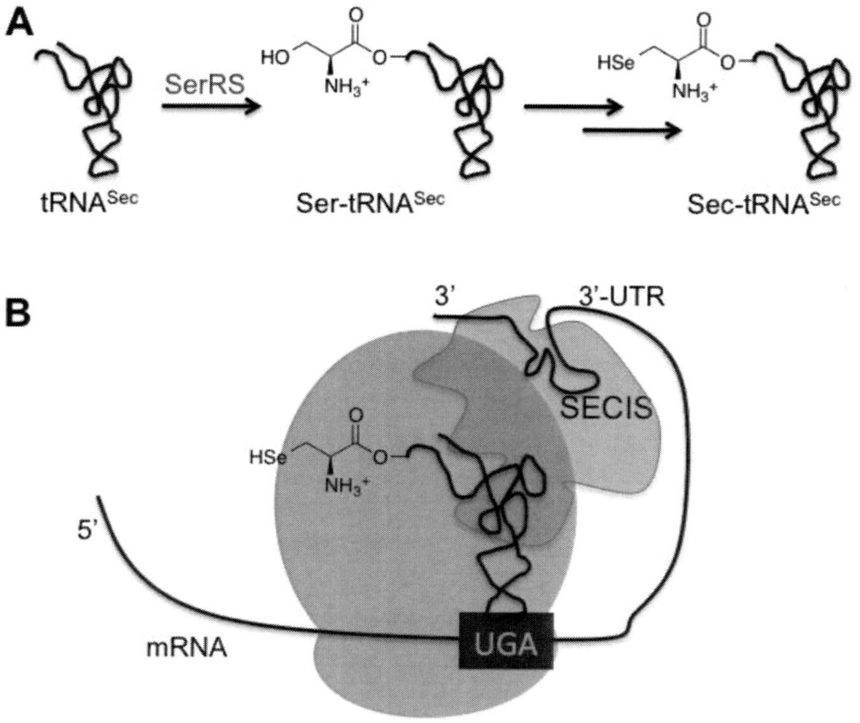

Figure 2. (A) Biological pathways to synthesize selenocysteyl-tRNA^{Sec} (Sec-tRNA^{Sec}). (B) Schematic representation of the mechanism of encoding selenocysteine in mammalian cells.

Another related unusual case is ribosomal frameshifting during protein synthesis [21]. Typically, proteins are synthesized based on a template mRNA with every three consecutive nucleotides being read as an amino acid. However, frameshifting occurs at low frequency: the ribosome slips by one base in either the 5' (-1) or 3' (+1) directions during translation. Frameshifting is related to nucleotide sequence, secondary structure and tertiary structure of an mRNA.

In the past decade, tremendous efforts have been put into investigation of molecular mechanisms related to ribosomal protein synthesis. Atomic structures of individual components involved in ribosomal protein synthesis have been elucidated. The 2009 Nobel Prize in Chemistry has been awarded to Venkatraman Ramakrishnan, Thomas A. Steitz and Ada E. Yonath for solving ribosome structure using X-ray crystallography. Detailed information on ribosomal protein synthesis is not the focus of this chapter, and can be found in other reviewer articles and chapters of this book [6-8].

2. Incorporation of Unnatural Amino Acids

The work to understand how proteins are synthesized has been very fruitful. In the meanwhile, researchers have developed methods to dramatically expand the repertoire of amino acids used in protein synthesis [22, 23]. Orthogonal tRNAs and aminoacyl synthetases have been engineered to encode unusual amino acids in response to nonsense codons and 4-

base codons. Additional translational machinaries including ribosome and translation factors have been mutated to increase the synthesis of unnatural proteins [24, 25]. Structurally and functionally manipulated proteins have been utilized to study biology and develop new therapeutics. Recent reviewers by us and others have summarized many details of this technology [22, 23]. Interested readers should refer to those indicated references. Here we only briefly describe the technology, link it with similar natural systems, focus on the re-engineering of components other than tRNAs and synthetases, and finally highlight its applications on therapeutics and vaccines.

Suppressor tRNAs for termination codons had been widely found in nature, so it was quite straightforward to propose a similar method to incorporate unnatural amino acids [11]. Initially, this was done *in vitro* using suppressor tRNAs pre-charged with unnatural amino acids, and *in vivo* by directly injecting charged tRNAs [26, 27]. Those charged tRNA molecules were made through either *in vitro* enzymatic reactions or methods that include organic synthesis. Research by Schultz and others established a procedure to genetically encode most components needed for incorporation of unusual amino acids (Figure 3) [28, 29]. The technology is often referred to as “genetic code expansion”, and has been widely adapted by the research community.

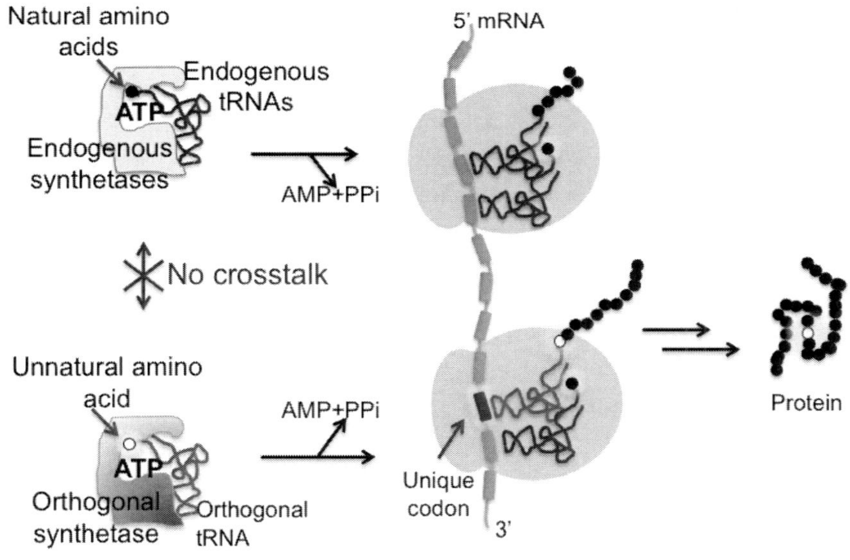

Figure 3. Schematic diagram of genetic encoding of unnatural amino acids in living cells.

In a typical experiment, a pre-engineered orthogonal tRNA with its anticodon complementary to a stop codon or a 4-based codon, and an also pre-engineered aminoacyl tRNA synthetase with preference toward the unnatural amino acid, are recombinantly expressed in cells. The unnatural amino acid is supplemented in the culture media. The resulting cells are capable to link the amino acid with the suppressor tRNA and synthesize modified proteins containing site-specifically inserted unnatural amino acids. This method is compatible with living cells, so it has become an indispensable tool for life science research. It is also an efficient and economical way to produce a large amount of nonnative proteins. Currently, the technology is available for genetic encoding of more than 90 unnatural amino acids harboring various reactive conjugation handles, photoactive functional groups, pre-installed post-translational modifications (PTMs), fluorophores, metal-chelating functional groups and other useful side chains [22, 23].

It is challenging to identify a pair of tRNA and synthetase orthogonal to cell endogenous pathways, and engineer them to gain selective activity toward a novel unnatural amino acid. In practice, orthogonal tRNA/synthetase pairs used in one organism are often derived from another organism in a different domain of life. For example, the tyrosyl tRNA and tyrosyl-tRNA synthetase pair from the archaeal *Methanocaldococcus jannaschii* (*Mj*TyrRS/*Mj*tRNA^{Tyr}) can be used in bacterial *E. coli* and *Mycobacterium tuberculosis* (MTB), while pairs derived from the *E. coli* tyrosyl tRNA and synthetase (*Ec*TyrRS/*Ec*tRNA^{Tyr}) have been used for genetic encoding of unnatural amino acids in eukaryotic cells [28, 29]. Many other important pairs for eukaryotic uses are derived from the *E. coli* leucyl tRNA and synthetase (*Ec*LeuRS/*Ec*tRNA^{Leu}). In addition, pyrrolysyl tRNAs and pyrrolysyl-tRNA synthetases (PylRS/tRNA^{Pyl}) from *Methanosarcina barkeri* and *Methanosarcina mazei*, are orthogonal in both prokaryotic and eukaryotic organisms, and have been engineered to encode many useful amino acids [30].

The anticodons of these suppressors have been switched so that they can pair with nonsense or 4-base codons. The first three bases of a 4-base codon need to be a less-used codon in the target organism (the corresponding endogenous tRNA is less abundant). In addition, wild-type synthetases have to be mutated to switch their substrate specificity from native amino acids to unnatural amino acids. Usually, rounds of positive and negative selections are performed. Briefly, synthetase libraries targeting at amino acid-binding residues are created by molecular biology. Both the tRNA and the synthetase mutants are imported into the organism cultured with media containing the supplemented unnatural amino acid. A gene necessary for cell survival under the given selection condition is induced for expression. However, nonsense or 4-base codons have been pre-inserted into its sequence. Only if a synthetase mutant can charge the tRNA with the unnatural amino acid to suppress nonsense or 4-base codons, cells would survive. Survivals from the positive selection will be subjected to a negative selection step, in which a toxic gene containing nonsense or 4-base codons will be expressed. No unnatural amino acid is provided in the negative selection step. Cells containing any synthetase mutant charging the tRNA with cell endogenous amino acids would be killed. The selection is often performed for multiple cycles to enrich synthetase mutants selective for the corresponding unnatural amino acid [23].

3. Engineering of Ribosome and Other Related Components

Suppression of nonsense and four-base codons is not very efficient. Recombinantly expressed and then charged orthogonal tRNAs have to compete with cell endogenous factors (Figure 4), i.e. translation termination factors (peptide release factors) or charged endogenous tRNAs that decode the first three bases of a four-base codon. Therefore, the yield of full-length proteins containing unnatural amino acids is often low. This problem is further amplified when multiple unusual codons are present in a single gene. Recent work has attempted to solve the problem by targeting individual or multiple steps involved in protein translation. For example, the interaction interface between the suppressor tRNA derived from *Mj*tRNA^{Tyr} and the *E. coli* elongation factor Tu (EF-Tu) has been re-engineered [31]. The improved tRNAs have been used to construct a series of pEvol plasmids showing robust amber suppression efficiency in *E. coli* cells [32]. We and others are currently performing similar work in yeast and mammalian cells to improved amber suppression in eukaryotic systems. Besides tRNAs and synthetases, other machinaries involved in protein translation, such as ribosome and other translational factors, have also been targeted. The purpose of those studies is to improve the efficiency of nonnative protein production, and/or enable the incorporation of unusual amino acids whose encoding is otherwise impossible.

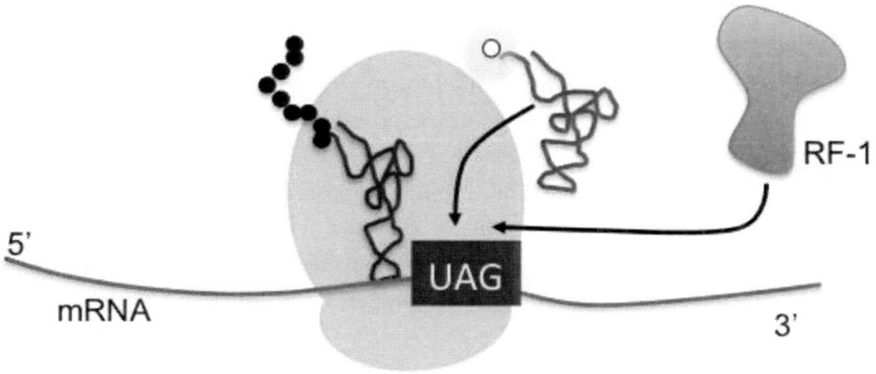

Figure 4. The competition between amber (TAG) codon suppression and RF-1 induced translation termination.

Elongation factors are critical enzymes involved in protein synthesize. Suppressor tRNAs carrying large nonnative amino acids are less tightly bound to elongation factor Tu (EF-Tu) than natural amino acids. Sisido *et al.* re-engineered the EF-Tu binding pocket for aminoacyl moieties of aminoacyl-tRNAs to increase its affinity toward large amino acids [33, 34]. Several bulk aromatic amino acids, which are hardly or only slightly incorporated by the wild-type EF-Tu, were successfully incorporated into proteins in the presence of the EF-Tu mutants.

Bacterial release factors (RFs) 1 and 2 catalyze translation termination at either UAG and UAA, or UAA and UGA, respectively (Figure 4). The large ribosomal subunit protein L11 is a highly conserved protein containing two domains, an N-terminal domain (L11N) and a C-terminal domain (L11C). L11 interacts with 23S rRNA and plays an important role in the RF1-mediated peptide release. L11C alone can also bind 23S rRNA. The ribosome, in which L11C is used to replace the full-length L11, shows translation efficiency comparable to the wild-type ribosome, but has lower efficiency in the RF1-mediated termination. Liu and his coworkers, therefore, overexpressed L11C in *E. coli* cell, to reduce RF1-mediated translation termination and increase amber suppression efficiency [35]. They demonstrated that three acetyllysine residues could be incorporated into a single peptide in a reasonable yield.

Sakamoto, Yokoyama and their coworkers engineered an *E. coli* strain, which lacks RF1 to terminate translation in response to UAG codons [36]. A few genetic modifications were, however, needed to circumvent the lethality of RF1 deletion. Several genes, which use UAG as their stop codons, were mutated. In their mutated strain, UAG was able to be assigned unambiguously to a natural or non-natural amino acid using different UAG-decoding tRNAs. They also demonstrated that *p*-iodophenylalanine could be incorporated in response to six in-frame amber codons in a model glutathione *S*-transferase (GST) protein. Similarly, Wang et al. also reported several RF1-deletion *E. coli* strains [37]. They found that R1 deletion could be tolerated by *E. coli*, as long as a certain version of RF2 is express in cells.[38] They confirmed that the critical residue in RF2 is Ala246. These reported *E. coli* strains are, undoubtedly, valuable tools for expression of proteins containing multiple unnatural amino acids at different residue sites.

To incorporate multiple chemically distinct unnatural amino acids into a single proteins, mutually orthogonal pairs that are also compatible with cell endogenous tRNAs, synthetases and amino acids are needed. First, Schultz and others reported the use of an *Mj*TyrRS/*Mj*tRNA^{Tyr} derived tRNA/synthetase pair and another pair derived from *Pyrococcus horikoshii* lysyl tRNA and synthetase in response to UAG and AGGA codons, respectively, for insertion of two different unnatural amino acids.(39) In addition, Liu et al. used *Mj*TyrRS/*Mj*tRNA^{Tyr} derived tRNA/synthetase pairs and PylRS/tRNA^{Pyl} derived pairs in the same *E. coli* cells to decode two nonsense codons (UAG and UAA). Chin and his coworkers, instead, reported the adaption of two orthogonal pairs directly from *Mj*TyrRS/*Mj*tRNA^{Tyr}, one pair responding to UAG and the other responding to AGGA.(40) Direct use of two nonsense codons, or one nonsense and one four-base codon, often leads to very low yield of protein production.

An exciting development is made by Chin and co-workers (Figure 5).(24) Orthogonal ribosomes were particularly developed for encoding unnatural amino acids. Briefly, a 16S rRNA library was built with mutations important for interactions at the ribosomal A site. The library was screened to identify mutants exhibiting a substantial increase in efficiency of decoding amber codons. Those mutant 16S rRNAs are likely to reduce the affinity between RF-1 and ribosome, so peptide releasing in response to UAG codons is reduced. Next, they engineered the ribosomal small subunit so that the mutated ribosome only binds a mutated SD sequence. These derived ribosomes can only translate exogenously introduced mRNAs, which harbor the mutated SD sequence. Endogenous mRNAs are excluded from the mutant ribosome due to the disrupted translation initiation. In the meanwhile, the synthesis of cell endogenous proteins is carried out by natural ribosomes.

More recently, Chin *et al.* further engineered an orthogonal ribosome for improved efficiency in decoding 4-base codons [25]. They showed that the mutant ribosome maintained its enhanced efficiency in decoding in-frame amber codons. Next, they used this orthogonal ribosome to synthesize proteins containing two different unnatural amino acids in response to both UAG and AGGA. One tRNA/synthetase pair was derived from *Mj*TyrRS/MjtRNA^{Tyr}, and another pair was derived from PylRS/tRNA^{Pyl}. They were able to generate a GST-calmodulin protein containing both azide and alkyne functional groups. The protein was subjected to click chemistry to build an intramolecular bridge through Cu(I)-catalyzed azide/alkyne Huisgen cycloaddition.

The research represents an interesting proof of concept that orthogonal ribosomes may be possibly re-engineered to reassign triplet and quadruplet codons. Research toward this direction is likely to establish biosynthetic pathways for polymers made with artificial building blocks.

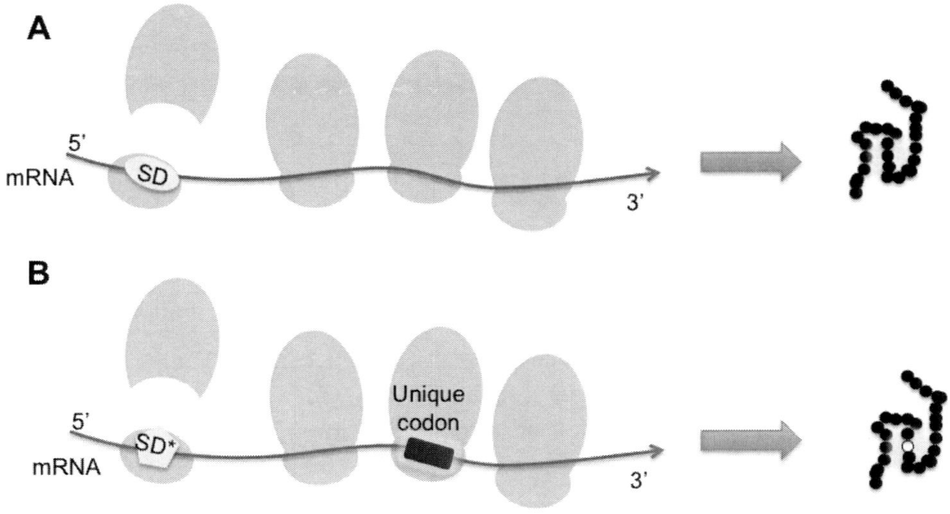

Figure 5. Protein synthesis in *E. coli* using (A) a wild-type ribosome and (B) an engineered orthogonal ribosome.

O-Phosphoserine (Sep) is an abundant posttranslational protein modification. Recently, Söll and coworkers reported a method to synthesize homogenous Sep-containing proteins in genetically modified *E. coli*.(41) Naturally, in some methanogenic archaea, there is no cysteinyl-tRNA synthetase. Instead, a Sep specific synthetase (SepRS) catalyzes the formation of the linkage between the amino acid *O*-phosphoserine and the corresponding cysteinyl-tRNA ($tRNA^{Cys}$). The *O*-phosphoserine charged $tRNA^{Cys}$ has low affinity with EF-Tu. It is subsequently converted to cysteine by the enzyme SepCysS in the presence of a sulfide donor.

Next, Cys-$tRNA^{Cys}$ is used by ribosome for protein synthesis. Söll et al. engineered a new amber suppressor from $tRNA^{Cys}$ by converting its anticodon to CUA (pair with UAG). An additional C20U mutation was made to improve the aminoacylation efficiency. It is worth noting that SepRS is not cross-reactive with any *E. coli* endogenous tRNA and can be overexpressed in *E. coli* cells. *E. coli* has a Sep-compatible transporter, so Sep was directly added to the growth medium.

The *E. coli* endogenous phosphoserine phosphatase gene, serB, was deleted to maintain adequate intracellular Sep concentration. Furthermore, a new EF-Tu was engineered and recombinantly expressed to increase its affinity. The engineered strain, which harbors a Sep-accepting transfer RNA, a cognate Sep-tRNA synthetase (SepRS), and an engineered EF-Tu (EF-Sep), was successfully utilized to synthesize the phosphorylated active form of human mitogen-activated ERK activating kinase 1 (MEK1). This research has built a new avenue to biosynthesize phosphoproteins for detailed studies of their biological properties.

To date, excluding tRNAs and synthetases, efforts to re-engineer protein synthesis-related components have been limited to *E. coli*. It remains to be determined whether similar strategies can be extended to eukaryotic (yeast and mammalian) cells and other industrial microbial strains for applications in biotechnology and pharmaceuticals.

4. Therapeutic Applications of Biosynthetic Unnatural Proteins

In this section, we focus on studies and applications related to therapy and vaccine development, where the production of homogenously modified protein in large quantities is desired.

The most dominant application is to produce protein conjugates containing toxins, radioisotopes and polyethylene glycols (PEG), and therapeutic protein-protein conjugates through site-specific chemical handles. The genetic code expansion technology is used to produce target proteins, which include an unnatural amino acid containing one of a few available bioorthogonal reactive chemical handles. The protein is then further functionalized *in vitro* with another small molecule or macromolecule through chemical reactions. The advantage is that it is highly specificity and can be used to label proteins containing multiple cysteines. The products are homogenous, so they can possibly be used as therapeutic agents due to their defined structure, pharmacological activity and toxicity. In one example, human growth hormone (hGH) containing *p*-acetylphenylalanine has been expressed and site-specifically PEGylated in large quantities.(42) The modified protein showed an enhanced serum half-life. The reagent is currently used in clinical trials. In other examples, antibodies have been efficiently expressed with ketone and alkyne functional groups for conjugation. For example, oligonucleotides have been site-specifically conjugated to antibodies, and the resulting oligobody molecules were used in immuno-PCR assays to detect Her2 positive cells.(43) In addition, toxins have been conjugated to anti-HER2 antibody trastuzumab through a *p*-acetylphenylalanine handle, resulting in targeted killing of HER2 positive cancer cells.(44) Furthermore, structurally defined bispecific antibodies were synthesized in multiple steps.(45) One antibody fragment that binds HER2 on breast cancer cells and another that binds proteins on immune killer T cells, were individually biosynthesized using genetically encoded unnatural amino acids with orthogonal chemical reactivity. Then a linker molecule was used to join the two to derive a bispecific antibody. In cell culture studies, the reagent at picomolar concentrations was able to bring the two cell types together, triggering T cell-induced cancer cell death.

Certain unnatural amino acids are known to increase the immune response of target proteins.(46, 47) They have been introduced into non-immunogenic or weak immunogenic proteins to break immunological tolerance, and used to generate therapeutic vaccines against self-proteins associated with cancer and inflammation. For example, murine TNF-α mutants containing *p*-nitrophenylalanine were injected into mice to induce high-titer antibodies that are cross-reactive with native murine TNF-α.(46) Similarly, sulfotyrosine and 3-nitrotyrosine have been incorporated into murine TNF-α and EGF, also triggering cross-reactive antibodies targeting wild-type proteins. Recent research has further investigated the mechanism, demonstrating that epitopes containing unnatural amino acids activate CD4 T cells through a neo-epitope recognition mechanism.(48) The T cell activation ultimately leads to activation of B cells that produce polyclonal antibodies against the wild-type protein. The mechanism also suggests that naturally occurring posttranslational modifications, such as nitration, may actually be linked to antibody-mediated autoimmune diseases.

Another important application is to site-specifically introduce unnatural amino acids containing therapeutic "warheads" into proteins. For example, the keto amino acid *p*-

benzoylphenylalanine was used to engineer cyclic peptides, which form covalent Schiff base adducts with the ε-amino group of Lys14 of the HIV protease.(49) Similarly, the amino acids containing boronic acid and acryl amide are being utilized to develop reagents reacting with cysteines of certain therapeutic targets. In another example, a recombinant sulfurated thrombin inhibitor hirudin (sulfohirudin) was made for enhanced affinity.(50) Sulfotysosine in this protein is critical for the tight interaction. Sulfohirudin was known as an anticoagulant leech protein, and the genetic code expansion technology has enabled its production in large quantities.

5. Future Directions

Biomolecular engineering of protein translation-related machinaries has now provided the ability to genetically encoding more than 90 unnatural amino acids. The early research was inspired directly by natural nonsense suppressors. Identification of orthogonal tRNA/synthetase pairs, including tyrosyl pairs and pyrrolysyl pairs, spurred the research filed. Further engineering on ribosome and translational factors improved and enhanced the technology for better yields and broader applications. However, most engineering still remains in *E. coli* cells. Further research is needed for yeast and mammalian cells, in which incorporation efficiency of unnatural amino acids is much lower. In addition, orthogonal ribosomes are very exciting, but it remains to be approved that global reassignment of codons, as either 3-base or 4-base codons, is possible. Analogous research in eukaryotic cells will further broaden the ability of this technology, which is expected to eventually help elucidate new biology and develop new therapeutics and vaccines.

Acknowledgment

We acknowledge support from the University of California, Riverside. We thank the Chinese American Faculty Association of South California (Robert T. Poe Faculty Development Grant to H.A.) and the UCR Academic Senate (Regents' Faculty Fellowship to H.A.).

References

[1] Crick F. Central Dogma of Molecular Biology. *Nature*.1970;227(5258):561-3.

[2] Ramakrishnan V. Ribosome Structure and the Mechanism of Translation. *Cell*. 2002;108(4):557-72.

[3] Chen H, Bjerknes M, Kumar R, Jay E. Determination of the optimal aligned spacing between the Shine-Dalgarno sequence and the translation initiation codon of *Escherichia coli* mRNAs. *Nucleic Acids Res.* 1994 Nov 25;22(23):4953-7.

[4] Preiss T, Hentze MW. Dual function of the messenger RNA cap structure in poly(A)-tail-promoted translation in yeast. *Nature*. 1998;392(6675):516-20.

[5] Frolova LY, Merkulova TI, Kisselev LL. Translation termination in eukaryotes: polypeptide release factor eRF1 is composed of functionally and structurally distinct domains. *RNA*. 2000;6(3):381-90.
[6] Korostelev A, Noller HF. The ribosome in focus: new structures bring new insights. T*rends Biochem. Sci.* 2007;32(9):434-41.
[7] Berk V, Cate JH. Insights into protein biosynthesis from structures of bacterial ribosomes. *Curr. Opin. Struct. Biol.* 2007;17(3):302-9.
[8] Schmeing TM, Ramakrishnan V. What recent ribosome structures have revealed about the mechanism of translation. *Nature*. 2009;461(7268):1234-42.
[9] Jukes TH, Osawa S. Evolutionary changes in the genetic code. *Comp. Biochem. Physiol. B*. 1993;106(3):489-94.
[10] Knight RD, Landweber LF, Yarus M. How mitochondria redefine the code. *J. Mol. Evol.* 2001;53(4-5):299-313.
[11] Murgola EJ. tRNA, suppression, and the code. *Annu. Rev. Genet.* 1985;19:57-80.
[12] Srinivasan G, James CM, Krzycki JA. Pyrrolysine encoded by UAG in *Archaea*: charging of a UAG-decoding specialized tRNA. *Science*. 2002;296(5572):1459-62.
[13] Hao B, Gong W, Ferguson TK, James CM, Krzycki JA, Chan MK. A new UAG-encoded residue in the structure of a methanogen methyltransferase. *Science*. 2002;296(5572):1462-6.
[14] Gaston MA, Zhang L, Green-Church KB, Krzycki JA. The complete biosynthesis of the genetically encoded amino acid pyrrolysine from lysine. *Nature*. 2011;471(7340):647-50.
[15] Cellitti SE, Ou W, Chiu H-P, Grunewald J, Jones DH, Hao X, et al. D-Ornithine coopts pyrrolysine biosynthesis to make and insert pyrroline-carboxy-lysine. *Nat. Chem. Biol.* 2011;7(8):528-30.
[16] Chambers I, Frampton J, Goldfarb P, Affara N, McBain W, Harrison PR. The structure of the mouse glutathione peroxidase gene: the selenocysteine in the active site is encoded by the 'termination' codon, TGA. *EMBO J.* 1986;5(6):1221-7.
[17] Bock A, Forchhammer K, Heider J, Leinfelder W, Sawers G, Veprek B, et al. Selenocysteine: the 21st amino acid. *Mol. Microbiol.* 1991;5(3):515-20.
[18] Copeland PR. Making sense of nonsense: the evolution of selenocysteine usage in proteins. *Genome Biol.* 2005;6(6):221.
[19] Yuan J, Palioura S, Salazar JC, Su D, O'Donoghue P, Hohn MJ, et al. RNA-dependent conversion of phosphoserine forms selenocysteine in eukaryotes and archaea. *Proc. Natl. Acad. Sci. USA*. 2006;103(50):18923-7.
[20] Berry MJ, Banu L, Harney JW, Larsen PR. Functional characterization of the eukaryotic SECIS elements which direct selenocysteine insertion at UGA codons. *EMBO J.* 1993;12(8):3315-22.
[21] Farabaugh PJ. Translational frameshifting: implications for the mechanism of translational frame maintenance. *Prog. Nucleic Acid Res. Mol. Biol.* 2000;64:131-70.
[22] Ai HW. Biochemical analysis with the expanded genetic lexicon. Anal. Bioanal. Chem. 2012;403(8):2089-102.
[23] Liu CC, Schultz PG. Adding new chemistries to the genetic code. *Annu. Rev. Biochem.* 2010;79:413-44.

[24] Wang K, Neumann H, Peak-Chew SY, Chin JW. Evolved orthogonal ribosomes enhance the efficiency of synthetic genetic code expansion. *Nat. Biotechnol.* 2007;25(7):770-7.

[25] Neumann H, Wang K, Davis L, Garcia-Alai M, Chin JW. Encoding multiple unnatural amino acids via evolution of a quadruplet-decoding ribosome. *Nature.* 2010;464(7287):441-4.

[26] Shimizu Y, Inoue A, Tomari Y, Suzuki T, Yokogawa T, Nishikawa K, et al. Cell-free translation reconstituted with purified components. *Nat. Biotech.* 2001;19(8):751-5.

[27] Saks ME, Sampson JR, Nowak MW, Kearney PC, Du F, Abelson JN, et al. An engineered Tetrahymena $tRNA_{Gln}$ for in vivo incorporation of unnatural amino acids into proteins by nonsense suppression. *J. Biol. Chem.* 1996;271(38):23169-75.

[28] Wang L, Brock A, Herberich B, Schultz PG. Expanding the genetic code of *Escherichia coli*. Science. 2001;292(5516):498-500.

[29] Chin JW, Cropp TA, Anderson JC, Mukherji M, Zhang Z, Schultz PG. An Expanded Eukaryotic Genetic Code. *Science.* 2003;301(5635):964-7.

[30] Chen PR, Groff D, Guo J, Ou W, Cellitti S, Geierstanger BH, et al. A facile system for encoding unnatural amino acids in mammalian cells. *Angew. Chem. Int. Ed.* 2009;48(22):4052-5.

[31] Guo J, Melancon CE, 3rd, Lee HS, Groff D, Schultz PG. Evolution of amber suppressor tRNAs for efficient bacterial production of proteins containing nonnatural amino acids. *Angew. Chem. Int. Ed.* 2009;48(48):9148-51.

[32] Young TS, Ahmad I, Yin JA, Schultz PG. An enhanced system for unnatural amino acid mutagenesis in *E. coli*. *J. Mol. Biol.* 2010;395(2):361-74.

[33] Nakata H, Ohtsuki T, Abe R, Hohsaka T, Sisido M. Binding efficiency of elongation factor Tu to tRNAs charged with nonnatural fluorescent amino acids. *Anal. Biochem.* 2006;348(2):321-3.

[34] Doi Y, Ohtsuki T, Shimizu Y, Ueda T, Sisido M. Elongation factor Tu mutants expand amino acid tolerance of protein biosynthesis system. *J. Am. Chem. Soc.* 2007;129(46):14458-62.

[35] Huang Y, Russell WK, Wan W, Pai PJ, Russell DH, Liu W. A convenient method for genetic incorporation of multiple noncanonical amino acids into one protein in *Escherichia coli*. *Mol. Biosyst.* 2010 Apr;6(4):683-6.

[36] Mukai T, Hayashi A, Iraha F, Sato A, Ohtake K, Yokoyama S, et al. Codon reassignment in the Escherichia coli genetic code. *Nucleic Acids. Res.* 2010;38(22):8188-95.

[37] Johnson DB, Xu J, Shen Z, Takimoto JK, Schultz MD, Schmitz RJ, et al. RF1 knockout allows ribosomal incorporation of unnatural amino acids at multiple sites. *Nat. Chem. Biol.* 2011;7(11):779-86.

[38] Johnson DB, Wang C, Xu J, Schultz MD, Schmitz RJ, Ecker JR, et al. Release Factor One Is Nonessential in *Escherichia coli*. *ACS Chem. Biol.* 2012;7(8):1337-44.

[39] Anderson JC, Wu N, Santoro SW, Lakshman V, King DS, Schultz PG. An expanded genetic code with a functional quadruplet codon. *Proc. Natl. Acad. Sci. USA.* 2004;101(20):7566-71.

[40] Neumann H, Slusarczyk AL, Chin JW. De novo generation of mutually orthogonal aminoacyl-tRNA synthetase/tRNA pairs. *J. Am. Chem. Soc.* 2010;132(7):2142-4.

[41] Park HS, Hohn MJ, Umehara T, Guo LT, Osborne EM, Benner J, et al. Expanding the genetic code of Escherichia coli with phosphoserine. *Science*. 2011;333(6046):1151-4.

[42] Cho H, Daniel T, Buechler YJ, Litzinger DC, Maio Z, Putnam AM, et al. Optimized clinical performance of growth hormone with an expanded genetic code. *Proc. Natl. Acad. Sci. USA*. 2011;108(22):9060-5.

[43] Kazane SA, Sok D, Cho EH, Uson ML, Kuhn P, Schultz PG, et al. Site-specific DNA-antibody conjugates for specific and sensitive immuno-PCR. *Proc. Natl. Acad. Sci. USA*. 2012;109(10):3731-6.

[44] Hutchins BM, Kazane SA, Staflin K, Forsyth JS, Felding-Habermann B, Smider VV, et al. Selective formation of covalent protein heterodimers with an unnatural amino acid. *Chem. Biol.* 2011;18(3):299-303.

[45] Kim CH, Axup JY, Dubrovska A, Kazane SA, Hutchins BA, Wold ED, et al. Synthesis of bispecific antibodies using genetically encoded unnatural amino acids. *J Am Chem Soc*. 2012;134(24):9918-21.

[46] Grunewald J, Tsao ML, Perera R, Dong L, Niessen F, Wen BG, et al. Immunochemical termination of self-tolerance. *Proc. Natl. Acad. Sci. USA*. 2008;105(32):11276-80.

[47] Grunewald J, Hunt GS, Dong L, Niessen F, Wen BG, Tsao ML, et al. Mechanistic studies of the immunochemical termination of self-tolerance with unnatural amino acids. *Proc. Natl. Acad. Sci. USA*. 2009;106(11):4337-42.

[48] Gauba V, Grunewald J, Gorney V, Deaton LM, Kang M, Bursulaya B, et al. Loss of CD4 T-cell-dependent tolerance to proteins with modified amino acids. *Proc. Natl. Acad. Sci. USA*. 2011;108(31):12821-6.

[49] Young TS, Young DD, Ahmad I, Louis JM, Benkovic SJ, Schultz PG. Evolution of cyclic peptide protease inhibitors. *Proc. Natl. Acad. Sci. USA*. 2011;108(27):11052-6.

[50] Liu CC, Schultz PG. Recombinant expression of selectively sulfated proteins in *Escherichia coli*. *Nat Biotechnol*. 2006;24(11):1436-40.

In: Ribosomes
Editors: Zhou Lin and Wang Liu

ISBN: 978-1-62417-698-2

Chapter 9

EMG1 Harbouring the Bowen-Conradi Syndrome D86G Substitution Localizes Normally but Has Reduced Stability

Joy Armistead and Barbara Triggs-Raine*
Department of Biochemistry and Medical Genetics, The University of Manitoba, Winnipeg, MB, Canada
The Manitoba Institute of Child Health, Winnipeg, MB, Canada

Abstract

Bowen-Conradi syndrome (BCS), a lethal autosomal recessive disorder which affects infants in the Hutterite population, is caused by a point mutation in *EMG1*, a highly conserved gene necessary for biogenesis of the small subunit of the ribosome. Research on EMG1 function has centered on yeast models, due to their genetic tractability for examining a cell-essential protein. The effects of the BCS-causing mutation on human EMG1 protein function are therefore unknown. We previously showed that EMG1 protein levels in BCS patient fibroblasts were severely reduced compared to control cells, and we therefore went on to examine EMG1 stability. Using pulse-chase metabolic labeling in cells transiently transfected with wild type or mutant EMG1, we found that the mutant protein rapidly associated with a detergent-insoluble fraction. Inhibition of the proteasome resulted in the accumulation of a higher-molecular weight band likely corresponding to ubiquitinated EMG1, suggesting that EMG1 is normally degraded by this pathway. Using immunofluorescence in fibroblasts, we found that both wild type and mutated EMG1 were localized to the nucleolus, and that proteasome inhibition resulted in the accumulation of EMG1 in nucleoplasmic foci which may correspond to sites of proteasome-mediated degradation. These data suggest that the mutated EMG1 protein is unstable, leading to a reduced level of EMG1 in the cell. Although the mutated protein localizes normally, the reduction in its level is likely to impact downstream pathways that significantly alter cell function in BCS patients.

* Corresponding author: Dr. Barbara Triggs-Raine, Dept. Biochemistry and Medical Genetics, The University of Manitoba, 745 Bannatyne Ave., Winnipeg, MB R3E OJ9, Phone: 204-789-3218, Fax: 204-789-3900, E-mail: traine@cc.umanitoba.ca.

Introduction

Bowen-Conradi syndrome (BCS) is a lethal autosomal recessive disorder which affects one in 355 live births in the Hutterite population. Children born with the disease are growth-delayed, and exhibit severe psychomotor retardation and failure to thrive, culminating in death at an average age of 13 months [1]. We determined that a c.257A>G; p.D86G mutation in the highly conserved gene *EMG1* (Essential for Mitotic Growth 1) is responsible for BCS. This mutation results in severely reduced levels of the EMG1 protein in patient fibroblasts . In yeast, studies of the function of Emg1 have shown that it is essential for the biogenesis of the small ribosomal subunit [3-5].

The biosynthesis of a mature, functional ribosome is an extremely complex process, involving the 79 proteins of the small and large subunit, hundreds of non-structural assembly proteins, dozens of small nucleolar RNAs, and the 5.8S, 18S, 28S, and 5S rRNAs. During ribosome biogenesis, a 45S rRNA precursor is transcribed, cleaved, folded and extensively modified to give rise to the mature 18S, 28S and 5S rRNAs. The most common rRNA alterations are methylation and pseudouridylation, which are thought to modify specificity, folding, and function in the context of the assembled ribosome [6-8]. Yeast Emg1 is necessary for the proper formation of the small subunit 18S rRNA [3-5,9,10].

The crystal structure of EMG1 orthologs in *Saccharomyces cerevisiae* and the archaebacterium *Methanocaldococcus janaschii* [11,12] revealed a distinctive core consisting of a five-stranded β-sheet and a deep trefoil knot in the C-terminal amino acid backbone, hallmarks of members of the superfamily of SPOUT-class methyltransferases [13]. SPOUT methyltransferases all mediate the methylation of the base or ribose moieties of RNA, and EMG1 was shown to catalyse the transfer of a methyl group from the donor, S-adenosylmethionine (SAM), to the hypermodified nucleotide N1-methyl-N3-(amino-3-carboxypropyl)-pseudouridine (m1acp3ψ) at position 1191 of yeast 18S rRNA (1248 in human 18S rRNA) [14]. The high degree of modification of m1acp3ψ may be an indication of its importance in the ribosome; in the mature small ribosomal subunit, this nucleotide is found at the P site in the decoding centre, and blocking pseudouridylation at this site in yeast reduces ribosomal fidelity [15]. There is evidence that yeast Emg1 may play other roles during ribosome biogenesis, including recruitment of the ribosomal protein RPS19 to the maturing ribosome, and removal of snR57, the snoRNA component of the small nucleolar ribonucleoprotein (snoRNP) responsible for 2'-O ribose methylation of G1570 in 18S rRNA [9]. Additionally, EMG1 has been proposed to play a structural role in the proper folding of 18S rRNA during biogenesis of the small subunit [16].

All characterized members of the SPOUT family exist as homodimers in solution. In the EMG1 homodimer, two parallel α-helices form the surface of the interface in one subunit and sit perpendicular to the two parallel α-helices in the other, forming a core four-helix bundle in the dimer stabilized by hydrophobic interactions [12]. The enzyme thus has two active sites and two binding sites for SAM, the methyl donor. Dimerization creates an extended, positively charged groove, which is thought to bind RNA [16]. The aspartate which is substituted by glycine in BCS patient cells, D86, is positioned near the end of alpha helix α2, one of the two parallel helices that form the interface between the two monomers. D86 binds and orients residue R84, which is necessary for RNA binding [11]. In addition, D86 is placed near the interface of the two dimer subunits. We predicted that substitution of aspartate for glycine would disrupt the alpha helix, collapsing the structure of the protein and revealing hydrophobic residues [2]. Thus, we expected the unstable mutated protein would aggregate to reduce the hydrophobic area exposed to the aqueous environment. We hypothesized that the loss of structure in D86G EMG1 would lead to its degradation by the proteasome, an important degradation pathway for misfolded proteins (reviewed in [17,18]). Proteins to be degraded undergo post-translational modification by ubiquitin, an 8.5 kDa protein [18-20]. Ubiquitin chains extended using the internal lysine 48 residue mark the protein to be sent to the 26S proteasome where it is deubiquitinated, unfolded, and degraded by proteolytic cleavage in an ATP-dependent manner (reviewed in [19,20]). Thus, inhibition of the proteasome would block EMG1 degradation and restore D86G EMG1 levels in BCS patient cells [21]. Proteasome inhibition has been shown to alter the mobility and localization of certain ribosomal proteins and processing factors [22], possibly because large complexes accumulate which cannot cycle through nucleolar subcompartments at normal rates. Consequently we also examined the sub-cellular localization of EMG1 before and after proteasome inhibition, hypothesizing that the predicted aggregates formed in BCS patient cells would alter the localization of EMG1. Mislocalization of ribosome-associated proteins has been connected with the ribosomopathies Treacher Collins syndrome [23] and Diamond-Blackfan anemia [24,25]. Moreover, both yeast and human GFP-tagged EMG1 bearing the BCS-causing mutation are mislocalized to the cytoplasm when they are overexpressed [14].

In order to examine the stability and localization of EMG1, we overexpressed the D86G and wild type (WT) EMG1 protein in baby hamster kidney (BHK) cells. The D86G substitution in EMG1 reduced the stability of the protein in comparison with wild type EMG1, leading to either aggregation or degradation via the proteasome pathway. The D86G substitution did not, however, alter EMG1 sub-cellular localization in BCS patient fibroblasts.

Materials and Methods

Study Subjects

Lymphoblast or fibroblast cell lines from patients were generated as part of a previous study to map the BCS gene [26]. Signed informed consent was obtained for all participants. This study was approved by the Health Research Ethics Board at the University of Manitoba. Anonymous control DNA samples were from non-Hutterite Manitobans and individuals from the Eastern United States. Control fibroblasts or lymphoblasts were from children of a similar age as BCS patients.

Cell Culture

BHK cells or fibroblasts were maintained in minimal essential medium (alpha modification) or Dulbecco's modified Eagle's medium (Invitrogen, Sigma, or Lonza). Lymphoblasts were maintained in RPMI-1640 medium (Invitrogen, Sigma). All media was supplemented with 10% fetal bovine serum, and 100 U penicillin and 100 μg streptomycin per ml (Invitrogen). Cells were maintained in a humidified chamber at 37°C in 5% CO_2.

Protein Isolation and Immunoblot Analysis

Soluble and insoluble proteins were isolated from cultured cells by lysis in a modified radio immunoprecipitation assay (RIPA) buffer (1% tert-octylphenoxy (poly) oxyethelene ethanol [IGEPAL]; 0.5% sodium deoxycholate; 0.1% SDS; 1.9 mM NaH_2PO_4; 8.1 mM Na_2HPO_4; 150 mM NaCl). RIPA-insoluble proteins were separated by centrifugation for 10 minutes in a microcentrifuge at 10000 x *g*, leaving the RIPA-soluble proteins in the supernatant. The pellet (insoluble fraction) was resuspended in RIPA buffer by 3 rounds of sonication for 5 seconds each. For isolation of cytoplasmic and nuclear fractions, cultured cells were lysed in Triton buffer (50 mM Tris pH 7.4; 1 mM EDTA; 150 mM NaCl; 1% Triton) and the fractions were separated by centrifugation at 2000 x *g* for 10 minutes in a microcentrifuge. The supernatant was removed and the pellet was sonicated for 3 rounds of 5 seconds each in nuclear lysis buffer (50 mM Tris pH 7.4; 10 mM EDTA; 1% SDS). All lysis buffers were supplemented 1 in 500 with a protease inhibitor cocktail (Sigma).

Protein concentrations were determined using either the Bradford assay (Bio-Rad) or the bicinchoninic acid assay (Thermo Scientific). Equal amounts of protein (20-25 μg) were separated at 160 V using 10% SDS-PAGE gels. Proteins were transferred to a nitrocellulose membrane in N-cyclohexyl-3-aminopropanesulfonic acid (CAPS) buffer at pH 11 (0.01 M CAPS, 10% methanol) at 100 V for one hour according to a modified method [27]. The membrane was blocked in 5% skim milk in Tris buffered saline Tween (TBST) (20 mM Tris pH 7.4; 0.15 M NaCl; 0.1% Tween) for one hour. Individual proteins were detected using a primary antibody diluted in 5% skim milk powder in TBST overnight at 4°C. The membrane was washed three times with TBST, incubated with horseradish peroxidase (HRP)-conjugated secondary antibody for one hour, washed, and visualized using Immobilon western chemiluminescent HRP substrate (Millipore). Primary antibody dilutions were used as follows: EMG1 rabbit polyclonal (Proteintech Group) at 1:2000, fibrillarin rabbit polyclonal (Aviva Systems Biology) at 1:5000, and ubiquitin rabbit polyclonal (StressMarq) at 1:1500. The secondary donkey anti-rabbit antibody (Jackson Immunoresearch) was diluted at 1:20000.

Pulse-Chase Metabolic Labeling of Protein

The generation of constructs expressing HA-tagged WT and D86G EMG1 was previously described [2]. Transfections were performed with 4 μg of plasmid DNA per 30 mm dish using Lipofectamine 2000 (Invitrogen) following the manufacturer's specifications. A vector expressing β-galactosidase (0.5 μg) was included as a transfection control. Metabolic labelling was performed following a standard method [28]. Transiently transfected BHK cells were starved in cysteine- (Cys) and methionine- (Met) free DMEM (Invitrogen) supplemented 1 in 100 with Glutamax (Invitrogen). Following starvation, nascent proteins were labelled by the addition of 200 μCi/ml ^{35}S Cys/Met EasyTag Expre^{35}S (Perkin Elmer) for a pulse period of 10 minutes. The label was then removed, the cells were washed twice in prewarmed unlabeled chase media (DMEM supplemented with 5 mM each of cysteine and methionine), then incubated in chase media for the indicated times. At the end of each chase period, cells were washed once with PBS, lysed in RIPA buffer, and separated into soluble and insoluble fractions. EMG1 was immunoprecipitated from the supernatant using a modified method [29]. Briefly, equal amounts of cell lysates were precleared by rotation with protein G beads (GE Healthcare) for one hour. Lysates were then transferred to new tubes containing protein G beads and a mixture of 2 μg polyclonal and 1 μg monoclonal anti-HA antibodies (BD Biosciences and Applied Biological Materials, respectively), and rotated overnight at 4°C. Protein G beads were pelleted by centrifugation, and the supernatant was removed and replaced with 0.5 ml wash buffer (0.5x RIPA or Triton buffer). Beads were washed four times, boiled in 30 μl of sample prep buffer (8% SDS, 250 mM Tris pH 6.8, 40% glycerol, 400 mM dithiothreitol, 0.08% bromophenol blue), and the supernatant was removed for loading onto a 10% SDS-PAGE gel. Sample loading for SDS-PAGE was normalized to protein concentration and corrected for transfection efficiency using β-galactosidase activity determined using a previously described protocol [30]. Following SDS-PAGE, the gel was fixed for 15 minutes in 10% acetic acid/ 30% methanol, and stained with Coomassie Blue to visualise proteins. After visualizing the proteins, the gel was incubated for 30 minutes in 1 M sodium salicylate/ 1% glycerol, and dried at 70°C. The dried gel was exposed to Kodak Biomax XAR or MS film overnight at -80°C with intensifying screens (Kodak), and the film was developed to visualise bands.

Proteasome Inhibition

The activity of the proteasome was inhibited by treating cells either with 10 μM of proteasome inhibitor I (Z-Ile-Glu(OtBu)-Ala-Leucinal) or 15 μM MG-132 (Cbz-Leu-Leu-Leucinal [Calbiochem]) dissolved in dimethylsulfoxide (DMSO), according to the manufacturer's recommendations. As a negative control, cells were treated with DMSO alone.

Immunofluorescence

Adherent cells were grown to sub-confluency on sterilized glass coverslips, rinsed with PBS, and fixed in 4% paraformaldehyde for 15 minutes at room temperature. The cover slips were washed in PBS 3 times for 5 minutes each, and quenched using 50 mM ammonium chloride in PBS for 10 minutes, followed by three PBS washes. The cell membrane was permeabilised with 0.1% Triton X-100 in PBS for 10 minutes, the coverslips were washed three times in PBS, and the cells were incubated in a blocking solution (3% bovine serum albumen, 0.05% Tween in PBS) for 60 minutes by placing 100 μl drops on Parafilm and inverting the coverslips on top. To prevent evaporation, coverslips were covered with culture plate lids and moist paper towels. The polyclonal EMG1 antibody (Proteintech Group) was diluted 1:500 in blocking solution and 100 μl was placed on Parafilm with the coverslip inverted on top for one hour. The Alexa Fluor 488-conjugated secondary antibody (Molecular Probes) was diluted 1:1000 and incubated on the coverslip for 30 minutes. The coverslips were rinsed twice with water, and incubated for 30 seconds in Hoechst 33342 (Invitrogen) diluted 1 μl in 100 ml of water to stain nuclei. Prolong Gold Antifade Reagent (Molecular Probes) was applied to a glass slide, and the coverslip inverted on top. After curing overnight, the slides were sealed with clear nail polish.

Results

EMG1 Levels Are Reduced in Lymphoblasts from BCS Patients

To confirm the reduction of EMG1 levels that we previously observed in BCS patient fibroblasts , lymphoblasts from normal controls and BCS patients were lysed and separated into nuclear and cytoplasmic fractions. Nuclear and cytoplasmic fractions, as well as total cell lysates, were separated by SDS-PAGE, and EMG1 was detected using an anti-EMG1 antibody. In both wild type and BCS patient samples, EMG1 was detected only in the nuclear fraction, and not in the cytoplasmic fraction. Similar to the fibroblasts, EMG1 levels in BCS patient lymphoblasts were strongly reduced compared to controls (Figure 1). EMG1 in the total cell lysate was so dilute that it was only detectable upon longer exposure (not shown). To verify the sub-cellular fractionation, the same membrane was stripped and re-probed to detect fibrillarin, a well-characterized nucleolar protein. As expected, fibrillarin was detected only in the nucleus. The fibrillarin antibody also confirmed equal loading of the WT and BCS samples.

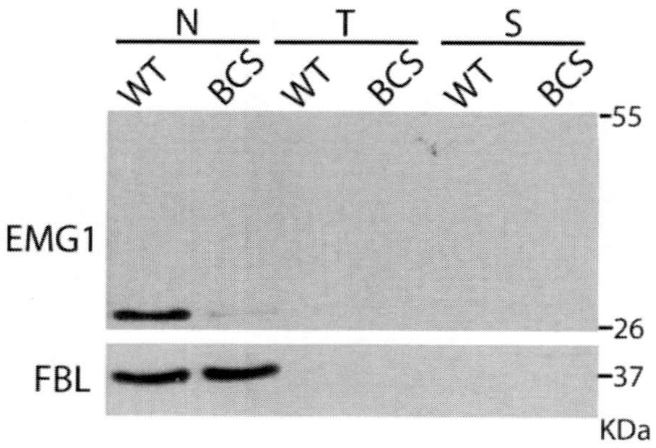

Figure 1. EMG1 is reduced in the nuclear fraction of Bowen-Conradi syndrome lymphoblasts. Nuclear fractions (N), total cell lysates (T) or post-nuclear supernatant fractions (S) of lymphoblasts from controls (WT) or BCS patients were evaluated for EMG1. EMG1 was only detected in the nuclear fraction, with reduced levels in BCS patient cells. The blot was reprobed for fibrillarin (FBL), a nucleolar protein, which served as a loading control in the nuclear fraction.

D86G EMG1 Rapidly Associates with an Insoluble Fraction

We previously showed in BHK cells transiently overexpressing HA-tagged WT and D86G EMG1 that the mutant protein was preferentially associated with a detergent-insoluble fraction [2]. These steady-state results were in accordance with our model of the human EMG1 structure which predicted that the D86G substitution would result in misfolding and subsequent aggregation of the protein, reducing its solubility. To further characterize the kinetics of D86G EMG1 protein reduction in the soluble fraction and association with an insoluble fraction, pulse-chase analysis of the EMG1 protein was performed in BHK cells. Cells transfected with empty vector were used as a control. Nascent HA-tagged EMG1 protein was labelled using ^{35}S cysteine/ methionine, and samples taken every two hours during the chase period were separated into RIPA-soluble and -insoluble fractions. In the soluble fraction, EMG1 was immunoprecipitated using an anti-HA antibody and SDS-PAGE gel loading was corrected for protein concentration and transfection efficiency. In the insoluble fraction, the high detergent concentration made it impossible to perform an immunoprecipitation and consequently, equal volumes were directly loaded onto the gel without further correction. Autoradiography showed strong bands at the expected molecular weight of 28 kDa (Figure 2). WT and D86G EMG1 started out at similar levels in the soluble fraction immediately following ^{35}S labelling (0 hours), but the D86G EMG1 levels diminished within two hours, in contrast to the WT EMG1 which was not reduced to a similar level until 24 hours. An examination of the insoluble fraction revealed the opposite results, with the mutant D86G EMG1 at much higher levels than WT at the 2 hour time point. Together, the detergent-soluble and -insoluble fractions show that D86G and WT EMG1 were produced at similar levels, but that the D86G EMG1 became rapidly insoluble, leaving very little soluble protein. These data support the hypothesis that the D86G EMG1 aggregates quickly following translation.

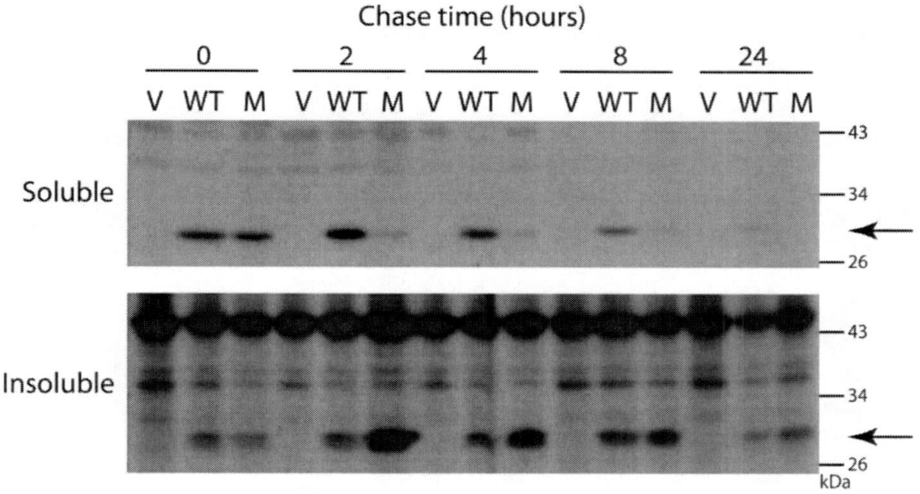

Figure 2. Metabolic labelling of EMG1 in BHK cells. BHK cells were transiently transfected with empty vector (V), wild type (WT), or mutant (M) EMG1-expressing plasmids. Following labelling of nascent proteins with ^{35}S cysteine and methionine, cell lysates were separated into detergent-soluble and insoluble fractions. EMG1 was immunoprecipitated in the soluble fraction and separated by gel electrophoresis, while aliquots from the insoluble fraction were directly loaded onto the gel. Immediately following labelling (0), WT and mutant EMG1 were present at equal levels in the soluble fraction, but levels of mutant protein were quickly reduced compared to WT protein. Conversely, levels of mutant EMG1 in the insoluble fraction increased quickly compared to WT protein.

EMG1 Is Degraded by the Proteasome Pathway

The low level of D86G EMG1 detected in BCS patient lymphoblasts and fibroblasts suggested that EMG1 is quickly degraded by the cell [18,22]. To determine if EMG1 was degraded via the proteasome pathway, proteasome activity was inhibited in fibroblasts using either 15 μM MG132 (Cbz-Leu-Leu-Leucinal) or 10 μM Proteasome Inhibitor I (PII) (Z-Ile-Glu(OtBu)-Ala-Leucinal). Successful proteasome inhibition was verified by the accumulation of ubiquitin-conjugated proteins, indicated by a high-molecular weight smear when nuclear fractions separated by SDS-PAGE were probed with an anti-ubiquitin antibody (Figure 3 A). Overnight treatment with MG132 consistently gave rise to a band at approximately 50 kDa that was detected by an anti-EMG1 antibody. This band was absent in untreated control cells and in cells treated with DMSO alone (arrow in Figure 3 B). This band may correspond to polyubiquitinated EMG1 as it has a molecular weight consistent with that of EMG1 plus three ubiquitin moieties (52.5 kDa). A band of a similar molecular weight appeared on immunoblots from cells treated overnight with PII (arrow in Figure 3 C). These data suggest that EMG1 is normally tagged by ubiquitin for degradation via the proteasome pathway. However, although proteasome inhibition strongly increased the putative polyubiquitinated EMG1 band, it did not restore D86G EMG1 to levels similar to WT levels.

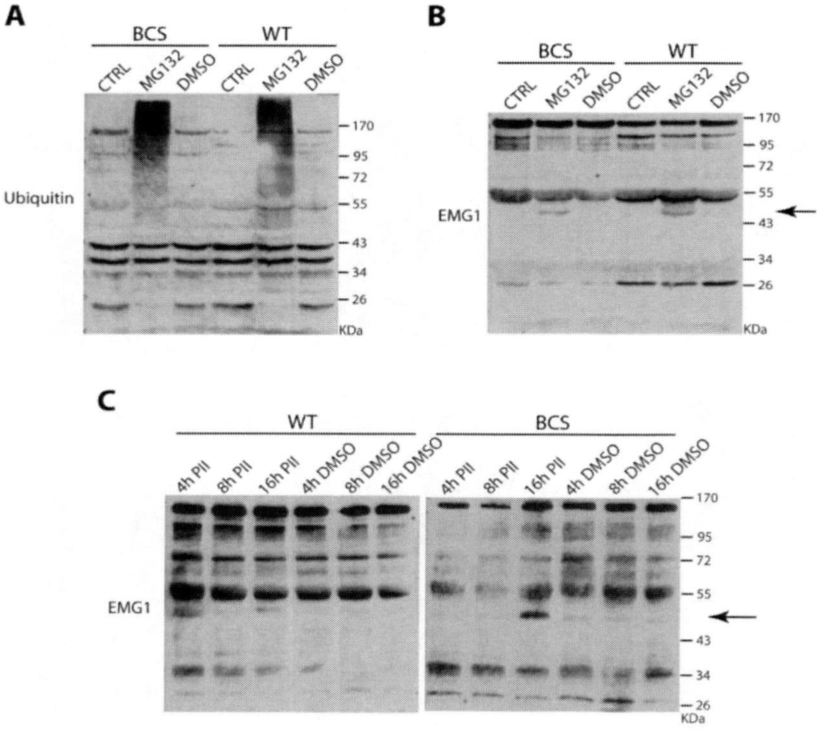

Figure 3. EMG1 is degraded by the proteasome pathway. (A) Cells were treated overnight with MG132, and nuclear lysates were separated by SDS-PAGE. The accumulation of proteins decorated with ubiquitin was detected by immunoblot, showing that treatment with MG132 did indeed inhibit the proteasome pathway. (B) In immunoblots probed for EMG1, a band at approximately 50 kDa was detected only in cells treated with MG132, but not in untreated cells, nor in cells treated with DMSO alone. (C) A 50 kDa band similar to that produced with MG132 treatment was also detected after treatment with PII, another proteasome inhibitor.

EMG1 is Localized in the Nucleoli in BCS Patient and Control Fibroblasts

To define the sub-cellular localization of EMG1, we examined EMG1 in cells by indirect immunofluorescence. Previous studies have suggested that the D86G substitution alters EMG1 sub-cellular localization [14], and we therefore compared control and BCS patient fibroblasts isolated from skin biopsies. Cells grown on coverslips were fixed and probed with an anti-EMG1 antibody, and a fluor-conjugated secondary antibody. In both the control and patient fibroblasts, EMG1 was found primarily in nucleoli, with a low level localized to nucleoplasmic puncta (Figure 4 A). There did not appear to be a difference in localization between the control and patient cells, and nucleoli were of similar number, size, and shape. The nucleolar sub-cellular localization is expected for a protein involved in ribosomal assembly, which takes place almost exclusively in the nucleolus.

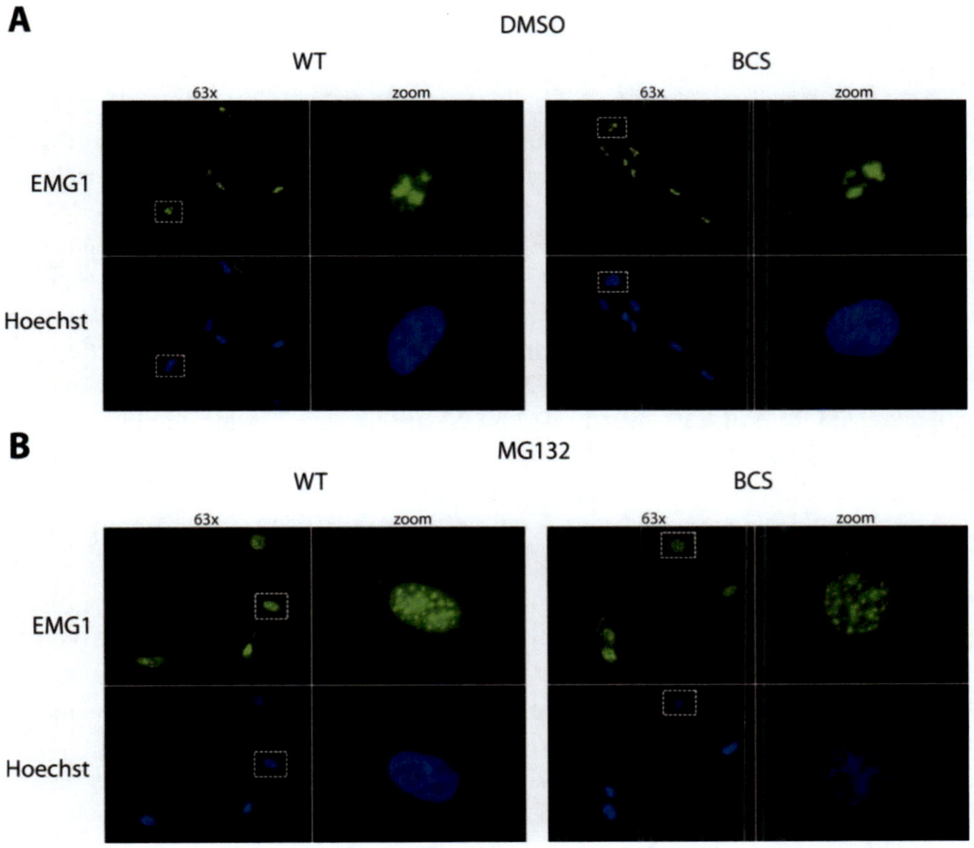

Figure 4. EMG1 is localized in fibroblast nucleoli at steady state, but migrates to nucleoplasmic puncta upon treatment with proteasome inhibitors. EMG1 (green) was detected in WT control and BCS patient fibroblasts by immunofluorescence. Nuclei (blue) are indicated by Hoechst 33342. The white box in the left panel indicates the area magnified in the right panel. (A) In untreated cells, EMG1 was found in nucleoli and scattered nucleoplasmic foci. (B) In cells treated overnight with MG132 to inhibit the proteasome pathway, EMG1 was localized to small nucleoplasmic puncta.

To observe if EMG1 sub-cellular localization was altered by proteasome inhibition, the protein was visualised in fibroblasts following overnight treatment with MG132. Proteasome inhibition led to a decrease in the nucleolar localization of EMG1 and an increase in its presence in nucleoplasmic foci (Figure 4 B). These foci may therefore correspond to nuclear sites of proteasome degradation, with EMG1 accumulating in the absence of proteasome activity. Again, there were no differences observed between control and patient cells.

Conclusion

All characterized members of the SPOUT family of methyltransferases exist as dimers in solution, with the extended surface created upon dimerization being necessary for binding the substrate RNA [13]. In fact, the catalytic residue is provided by the opposite subunit [13], and thus the dimeric structure is intimately linked with its enzyme function. In EMG1, a member of the SPOUT methyltransferases, the substitution of aspartate 86 for a glycine in Bowen-Conradi syndrome is predicted to expose hydrophobic residues, reducing protein stability [2]. This was evidenced by a reduction in D86G EMG1 levels at steady state in BCS patient fibroblasts and lymphoblasts (Figure 1). We therefore examined EMG1 stability using pulse-chase metabolic labeling. In BHK cells overexpressing EMG1, we found that WT and D86G EMG1 were initially present at similar levels in the soluble fraction of the cell, but that D86G EMG1 levels were rapidly reduced to a fraction of WT levels (Figure 2). Conversely, D86G EMG1 in the insoluble fraction accumulated quickly in comparison with WT EMG1. It is likely that the overexpressed D86G EMG1 misfolded and formed aggregates when hydrophobic residues were exposed to the surface, leaving less D86G EMG1 protein in a soluble state. This supports our previous work which suggested that the defect caused by the mutation in *EMG1* takes place at the protein level [2].

An examination of the mechanism by which EMG1 is degraded suggested that a polyubiquitinated form of EMG1, indicated by the appearance of a higher molecular weight band, accumulated upon treatment with MG132 or PII (Figure 3). This suggests that EMG1 is normally degraded by the proteasome pathway. In a study of Diamond-Blackfan anemia, proteasome inhibition increased the levels of mutated RPS19 protein [31]. However, proteasome inhibition did not restore D86G EMG1 to normal levels, indicating that a proportion of the D86G EMG1 may also aggregate. Degradation by the proteasome is in competition with aggregation for misfolded substrates [17] and consequently the proteasome must degrade misfolded proteins before they have a chance to aggregate. Therefore, it is possible that the majority of D86G EMG1 aggregates following proteasome inhibition, rendering it undetectable in the nuclear fraction.

At steady state, immunofluorescence showed that both WT and D86G EMG1 were found in nucleoli and in scattered nuclear puncta of fibroblasts (Figure 4 A). Following proteasome inhibition, EMG1 was re-localized from nucleoli and into nucleoplasmic foci (Figure 4 B). Due to the accumulation of EMG1 in these puncta after MG132 treatment, it is possible that they contain proteasomes to which EMG1 is transported for degradation. Our immunofluorescence results did not indicate any difference in localization between WT and D86G EMG1. This is contrary to results published by Meyer *et al* in 2011, who showed that GFP-tagged D86G EMG1 in HEK293T cells and D90G Emg1 (the yeast equivalent of the D86G mutation) in yeast were mislocalized to the nucleoplasm and the cytoplasm. These proteins were overexpressed and under the control of a tetracycline-inducible promoter, and it is possible that the overexpression of EMG1 in these experiments led to its mislocalization.

Our experiments detected endogenous D86G EMG1 in control and BCS patient fibroblasts, using an anti-EMG1 antibody. It seems more likely that the defects seen in BCS patients are the result of insufficient EMG1 levels in proliferating cells, rather than of its mislocalization.

Because ribosome biogenesis takes place primarily in the nucleolus, many studies of the molecular causes of ribosomopathies have focussed on sub-cellular localization of the mutated proteins, with variable results. In a study identifying nucleolar targeting sequences in RPS19, Da Costa et al (2003) examined two Diamond-Blackfan anemia-causing mutations which had the potential to negatively affect sub-cellular localization of the protein. When GFP fusion proteins bearing these substitutions were expressed in COS-7 cells, distribution of the protein was altered from the nucleolus to the nucleoplasm or cytoplasm. Notably, several other disease-associated mutations did not alter RPS19 nucleolar localization [24,25,31]. Similar studies of mutations in treacle, the protein associated with Treacher Collins syndrome, showed mislocalization to the nucleoplasm [23]. However, SBDS proteins carrying different mutations retained nucleolar localization in Shwachman-Diamond syndrome patient cells [32]. Similarly, our studies have shown that the D86G substitution alone is insufficient to alter the localization of EMG1. Taken together, these results indicate that mislocalization cannot be considered a common theme in ribosomopathies.

In sum, our results indicate that the D86G EMG1 protein which is responsible for BCS is unstable, causing a rapid reduction of soluble EMG1 levels following translation. While the wild type EMG1 is normally degraded by the proteasome pathway, D86G EMG1 also forms aggregates due to the exposure of hydrophobic residues. Because EMG1 aids in the biogenesis of the small ribosomal subunit [3-5,9,10], the reduction in available EMG1 protein in the cells of BCS patients may lead to a reduction in ribosome levels, negatively affecting cell proliferation during development.

Acknowledgments

This study was funded by grants from the Canadian Institutes of Health Research MOP 62786 (B. T.-R.) and the Manitoba Institute of Child Health. J.A. was supported by studentships from the Canadian Institutes of Health Research, the Manitoba Institute of Child Health and the Manitoba Health Research Council.

References

[1] Lowry RB, Innes AM, Bernier FP, McLeod DR, Greenberg CR, Chudley AE, et al. Bowen-Conradi syndrome: a clinical and genetic study. *Am J Med Genet A* 2003 Jul 30;120A(3):423-428.

[2] Armistead J, Khatkar S, Meyer B, Mark BL, Patel N, Coghlan G, et al. Mutation of a gene essential for ribosome biogenesis, EMG1, causes Bowen-Conradi syndrome. *Am J Hum* Genet 2009 Jun;84(6):728-739.

[3] Liu PC, Thiele DJ. Novel stress-responsive genes EMG1 and NOP14 encode conserved, interacting proteins required for 40S ribosome biogenesis. *Mol Biol Cell* 2001 Nov;12(11):3644-3657.

[4] Eschrich D, Buchhaupt M, Kotter P, Entian KD. Nep1p (Emg1p), a novel protein conserved in eukaryotes and archaea, is involved in ribosome biogenesis. *Curr Genet* 2002 Feb;40(5):326-338.

[5] Bernstein KA, Gallagher JE, Mitchell BM, Granneman S, Baserga SJ. The small-subunit processome is a ribosome assembly intermediate. *Eukaryot Cell* 2004 Dec;3(6):1619-1626.

[6] Eichler DC, Craig N. Processing of eukaryotic ribosomal RNA. *Prog Nucleic Acid Res Mol Biol* 1994;49:197-239.

[7] Fromont-Racine M, Senger B, Saveanu C, Fasiolo F. Ribosome assembly in eukaryotes. *Gene* 2003 Aug 14;313:17-42.

[8] Henras AK, Soudet J, Gerus M, Lebaron S, Caizergues-Ferrer M, Mougin A, et al. The post-transcriptional steps of eukaryotic ribosome biogenesis. *Cell Mol Life Sci* 2008 Aug;65(15):2334-2359.

[9] Buchhaupt M, Meyer B, Kotter P, Entian KD. Genetic evidence for 18S rRNA binding and an Rps19p assembly function of yeast nucleolar protein Nep1p. *Mol Genet Genomics* 2006 Sep;276(3):273-284.

[10] Wurm JP, Meyer B, Bahr U, Held M, Frolow O, Kotter P, et al. The ribosome assembly factor Nep1 responsible for Bowen-Conradi syndrome is a pseudouridine-N1-specific methyltransferase. *Nucleic Acids Res* 2010 Jan 4.

[11] Leulliot N, Bohnsack MT, Graille M, Tollervey D, Van Tilbeurgh H. The yeast ribosome synthesis factor Emg1 is a novel member of the superfamily of alpha/beta knot fold methyltransferases. *Nucleic Acids Res* 2008 Feb;36(2):629-639.

[12] Taylor AB, Meyer B, Leal BZ, Kotter P, Schirf V, Demeler B, et al. The crystal structure of Nep1 reveals an extended SPOUT-class methyltransferase fold and a pre-organized SAM-binding site. *Nucleic Acids Res* 2008 Mar;36(5):1542-1554.

[13] Tkaczuk KL, Dunin-Horkawicz S, Purta E, Bujnicki JM. Structural and evolutionary bioinformatics of the SPOUT superfamily of methyltransferases. *BMC Bioinformatics* 2007 Mar 5;8:73.

[14] Meyer B, Wurm JP, Kotter P, Leisegang MS, Schilling V, Buchhaupt M, et al. The Bowen-Conradi syndrome protein Nep1 (Emg1) has a dual role in eukaryotic ribosome biogenesis, as an essential assembly factor and in the methylation of Psi1191 in yeast 18S rRNA. *Nucleic Acids Res* 2011 Mar;39(4):1526-1537.

[15] Baudin-Baillieu A, Fabret C, Liang XH, Piekna-Przybylska D, Fournier MJ, Rousset JP. Nucleotide modifications in three functionally important regions of the Saccharomyces cerevisiae ribosome affect translation accuracy. *Nucleic Acids Res* 2009 Dec;37(22):7665-7677.

[16] Thomas SR, Keller CA, Szyk A, Cannon JR, Laronde-Leblanc NA. Structural insight into the functional mechanism of Nep1/Emg1 N1-specific pseudouridine methyltransferase in ribosome biogenesis. *Nucleic Acids Res* 2011 Mar;39(6):2445-2457.

[17] Kopito RR. Aggresomes, inclusion bodies and protein aggregation. *Trends Cell Biol* 2000 Dec;10(12):524-530.

[18] Tyedmers J, Mogk A, Bukau B. Cellular strategies for controlling protein aggregation. *Nat Rev Mol Cell Biol* 2010 Nov;11(11):777-788.

[19] Tai HC, Schuman EM. Ubiquitin, the proteasome and protein degradation in neuronal function and dysfunction. *Nat Rev Neurosci* 2008 Nov;9(11):826-838.

[20] Jadhav T, Wooten MW. Defining an embedded code for protein ubiquitination. *J Proteomics Bioinform* 2009 Jul 24;2:316.

[21] Lee DH, Goldberg AL. Selective inhibitors of the proteasome-dependent and vacuolar pathways of protein degradation in Saccharomyces cerevisiae. *J Biol Chem* 1996 Nov 1;271(44):27280-27284.

[22] Stavreva DA, Kawasaki M, Dundr M, Koberna K, Muller WG, Tsujimura-Takahashi T, et al. Potential roles for ubiquitin and the proteasome during ribosome biogenesis. *Mol Cell Biol* 2006 Jul;26(13):5131-5145.

[23] Marsh KL, Dixon J, Dixon MJ. Mutations in the Treacher Collins syndrome gene lead to mislocalization of the nucleolar protein treacle. *Hum Mol Genet* 1998 Oct;7(11):1795-1800.

[24] Da Costa L, Tchernia G, Gascard P, Lo A, Meerpohl J, Niemeyer C, et al. Nucleolar localization of RPS19 protein in normal cells and mislocalization due to mutations in the nucleolar localization signals in 2 Diamond-Blackfan anemia patients: potential insights into pathophysiology. *Blood* 2003 Jun 15;101(12):5039-5045.

[25] Angelini M, Cannata S, Mercaldo V, Gibello L, Santoro C, Dianzani I, et al. Missense mutations associated with Diamond-Blackfan anemia affect the assembly of ribosomal protein S19 into the ribosome. *Hum Mol Genet* 2007 Jul 15;16(14):1720-1727.

[26] Lamont RE, Loredo-Osti J, Roslin NM, Mauthe J, Coghlan G, Nylen E, et al. A locus for Bowen-Conradi syndrome maps to chromosome region 12p13.3. *Am J Med Genet A* 2005 Jan 15;132A(2):136-143.

[27] Towbin H, Staehelin T, Gordon J. Electrophoretic transfer of proteins from polyacrylamide gels to nitrocellulose sheets: procedure and some applications. 1979. *Biotechnology* 1992;24:145-149.

[28] Bonifacino JS. Metabolic labeling with amino acids. *Curr Protoc Protein Sci* 2001 May;Chapter 3:Unit 3.7.

[29] Bonifacino JS, Dell'Angelica EC, Springer TA. Immunoprecipitation. *Curr Protoc Protein Sci* 2001 May;Chapter 9:Unit 9.8.

[30] Rosenthal N. Identification of regulatory elements of cloned genes with functional assays. *Methods Enzymol* 1987;152:704-720.

[31] Cretien A, Hurtaud C, Moniz H, Proust A, Marie I, Wagner-Ballon O, et al. Study of the effects of proteasome inhibitors on ribosomal protein S19 (RPS19) mutants, identified in patients with Diamond-Blackfan anemia. *Haematologica* 2008 Nov;93(11):1627-1634.

[32] Austin KM, Leary RJ, Shimamura A. The Shwachman-Diamond SBDS protein localizes to the nucleolus. *Blood* 2005 Aug 15;106(4):1253-1258.

Index

A

B

C

D

E

I

K

L

M

N

O

P

T

U

V

W

Y